Point defects in
group IV semiconductors
common structural and physico-chemical aspects

by
S. Pizzini

A self-consistent, microscopic model of individual- and -reacted point defects requires a reliable connection with the experimentally deduced structural, spectroscopic and thermodynamic properties of the defect centres, to allow their unambiguous identification.

Aim of this book is to focus on the properties of defects in semiconductors of the fourth group under a physico-chemical approach, capable to demonstrate whether the full acknowledgement of their chemical nature could account for several problems encountered in practice or would suggest further experimental or theoretical accomplishments.

It will be shown how difficult the fulfilment of self-consistency conditions can be, even today, after more than four decades of dedicated research work, especially in the case of compound semiconductors (SiC in this book), but also in the apparently simplest cases of silicon and germanium also because microscopic models do not account, jet, for defect interactions in real solids.

Keywords

Point Defects in Silicon, Point Defects in Germanium, Point Defects in Diamond, Point Defects in Silicon Carbides, Point Defect-Impurity Complexes, Defect Modeling, Self-Diffusion, Impurity Diffusion

Cover Graphic

The carbon vacancy-silicon antisite complex. After A. Mattausch [197]. With kind permission from Alexander Matthausch and Professor Michel Bockstedte, Erlangen-Nuremberg University.

Point defects in
group IV semiconductors
common structural and
physico-chemical aspects

written by

S. Pizzini

Department of Materials Science, University of Milano-Bicocca, Italy

sergio.pizzini@mpvresearch.eu, sergio.pizzini@unimib.it

Published by **Materials Research Forum LLC**
Millersville, PA 17551, USA

Published as part of the book series
Materials Research Foundations
Volume 10 (2017)
ISSN 2471-8890 (Print)
ISSN 2471-8904 (Online)

Print ISBN 978-1-945291-22-7
ePDF ISBN 978-1-945291-23-4

Distributed worldwide by

Materials Research Forum LLC
105 Springdale Lane
Millersville, PA 17551
USA
http://www.mrforum.com

Manufactured in the United State of America
10 9 8 7 6 5 4 3 2 1

Table of Contents

Preface

Chemically pure and structurally perfect solids are unattainable as the products of a chemical synthesis or of a growth process from a melt or a gaseous phase, since they would present intrinsic thermodynamic instability.

Thermodynamic equilibrium conditions are attained with the spontaneous, or mechanically activated, formation of point and extended defects and with a spontaneous impurity contamination from the environment, in both geosynthetic solids, like diamond, or in synthetic single-crystals or polycrystalline semiconductors of the fourth group, the materials under discussion in this book.

Point-defects and impurities do behave as chemical species in thermodynamic equilibrium when the *system* where they sit-in is in thermal and chemical equilibrium with the environment and is not subject to external interactions of other nature, as energetic radiations or particle- and electron- beams, capable to modify the thermal equilibrium conditions.

Here the *system* is a portion of a single crystal, or a homo-or heteroepitaxial layer on a convenient substrate, with a well-known preparation history and a well-known chemical composition.

Under the previously defined conditions, a generic semiconducting crystal is a multinary phase, consisting of all the chemical or chemically-equivalent, species present (lattice atoms, defects, impurities[1]), and the Gibbs free energy of the phase, at constant temperature, is a function of all the possible chemical interactions occurring into the system.

Thus, the equilibrium properties of defects and impurities, and their equilibrium concentrations, in an elemental semiconductor, or in a stoichiometric or non-stoichiometric compound semiconductor, are those of these species in a multinary solution, whose energy depends on several pairwise interactions occurring among the species present or on the stoichiometric deviations.

Since both defects and impurities are, generally, present in extremely dilute conditions, ideal solution conditions could be, often, expected for semiconductor systems, unless stable defect-defect or defect-impurity pairs could be formed, because of large values of pair formation energy, as is the case of vacancies, that behave as rather chemically active species.

---

[1] Extended defects are non-equilibrium defects

We will see that the properties of isolated or of some reacted-defects in thermal equilibrium might be modelled with *ab initio* computations and experimentally investigated with several spectroscopic methods and with self-diffusion and impurity-diffusion measurements, these last providing a proper mean to determine their effective concentrations and their formation energies, though under severe experimental and theoretical difficulties, about which will be discussed in details.

In the best case, the experimentally detected defect properties are the *true* equilibrium properties of defects relative to the specific system investigated, but not directly extendable to a different physical system, nor to a model system used for computational work, that consists of a limited portion of an ideally pure and crystalline lattice in which a defect is suitably introduced.

The main attempt of this Book is to discuss the properties of defects in semiconductors of the fourth group under a physico-chemical approach, capable to demonstrate whether the full acknowledgement of their chemical nature would account for several problems encountered in practice or would suggest further experimental or theoretical accomplishments.

Chapter 1

*...We have to work with a material that is never completely pure, and which always contains a range of defcts although a particular one may be predominant. We shall find that impurities can cluster or precipitate and this process leads to structural defects. The latter defects may then act as nucleation sites for further interactions with impurities. It follows that the individual types of defects listed above are in no sense independent of each other. ..... R. C. Newman, 1982, Defects in Silicon, Rep.Progr.Phys. **45***

1.   Introduction

The best introduction to this Book is the sentence of Newman written in 1982 and reported above, since it pays full account for the role of defect interactions in semiconductors. As materials are never entirely pure, defects are chemical entities not independent of each other.

Since then, the technological development of microelectronics and optoelectronics continued to be seriously influenced by the presence of point- and extended-defects in silicon and, later, in elemental and compounds semiconductors of the group IV and III-V, which are the materials on which relies, today, the progress of modern microelectronics and optoelectronics.

Defects, on one side, might behave as life-time killers, thus severely degrading the behaviour of devices, but, on the other side, uniquely account for mass- and charge-transport processes in semiconductors, with tremendous influence on the performances of processes and devices.

Like subatomic particles, needed for the foundation of a whole theory of the Universe, point defects in ionic-, metallic- and semiconductor-crystals remain exotic particles, whose properties are well predicted by theory, but whose presence is often difficult to be experimentally detected in thermodynamic equilibrium conditions as individual species.

Therefore, in analogy with subatomic particles that are studied in modern hadronic colliders under impact of GeV protons or electron beams, defects in semiconductors were and still are investigated by irradiating semiconductor crystals with electrons, X-and γ-Ray beams[2], taking their energies above the displacement threshold energy of the atoms in the lattice. The nature and properties of defects subsequently generated, primarily consisting of vacancies and self-interstitials, are considered equivalent to those of defects thermally generated in thermodynamic equilibrium conditions.

Solid state physics, quantum mechanics, and physical chemistry are the conceptual and experimental frames on which was, and still is based the development of point defect models and the experimental detection of point defects. Here, chemistry not only accounts for the experimental deviations from a purely physical model of point defects, but addresses at their true chemical nature in covalent semiconductors.

As a first example of this intimacy, physico chemical concepts based on ionic solution theory were adopted in the early thirties of the last century by W. Schottky and C. Wagner[3] [1-2] to explain the electrical conductivity in ionic solids, conceived as due to point defects (vacancies and self- interstitials) that assist the charge carrier's migration in the ionic lattice.

Fast ionic solid conductivity found in the eighties of the last century theoretical accomplishments by considering an interplay of fast ionic migration and open structure details. The same concept, in an apparently very different field, is adopted to model the electrical conductivity of ice VII at 10 GPa, a phase of ice stable at pressures higher than 2 GPa, with its bcc oxygen lattice and a random hydrogen bonds network, that is shown to favor the ability of protons to flow through the oxygen lattice, with an exceptionally high diffusion coefficient [3], resembling the Ag^+ flow in superionic solids, like Ag_4RbI_5 [4].

---

[2] Irradiation with ions presents the disadvantage of ion-defect interactions.

[3] The author had the chance to invite Prof. Wagner at a Symposium held in Italy on fast ion transport in solids, few years before he died.

The presence of vacancies and self-interstitials in diamond was deduced by Coulson [5], in the early days of the semiconductor science, from the optical absorptions between 1.65 and 2.4 eV, with a sharpest peak at 1.67 eV (the GR1 band) detected in irradiated diamond. He thereafter suggested that the GR1 band, but also the other known optical bands, should be due to electronic transitions coupled to vibrational motion of defects (isolated vacancies, isolated self-interstitials and to simple (chemical) combinations of them) using molecular chemistry concepts to enlighten the structure and properties of vacancies and anticipating of at least three decades the modern views on defect complexes.

He assumed also that *in the immediate neighborhood of the vacancy the electronic properties are depending just from four unpaired electrons[4]*, a concept which we still share today.

Just only twenty years later, the presence of vacancies in silicon irradiated with 1.5 MeV electrons was deduced from their Electron Spin Resonance (ESR) signatures by Watkins [6], who gave also the first microscopic model of the vacancy and suggested that vacancies and self-interstitials are the elementary lattice excitations.

In the following years, with the increasing amount of information available about defect generation and interaction processes, mass action law concepts and equilibrium and non-equilibrium thermodynamics have been systematically used to implement defect engineering strategies, as is the case of gettering and hydrogen passivation techniques, or of processes addressed at the suppression of defect activity in semiconductor technology [7]. As well, the massive use of *ab initio*, Molecular Dynamics and Density Functional Theories (DFT) enabled the tremendous progress of our theoretical defect knowledge.

Eventually, chemistry accounts well for defect interaction and defect complex formation, an issue difficult to be modelled and formalized without the use of chemical concepts, and, last but not least, most of the spectroscopic techniques and of the theoretical methods used today for the numerical evaluation of defect properties have a physico-chemical origin. This is the case of ESR (Electron Spin Resonance), FTIR- (Fourier Transform Infra-Red) and Raman-spectroscopies, as well as of the Density Functional Theory (DFT), which has a major role in the theoretical studies carried- out on defects.

A self-consistent, microscopic model of individual- and -reacted point defects requires a reliable connection with the experimentally deduced structural, spectroscopic and thermodynamic properties of the defect centres, to allow their unambiguous identification.

---

[4] Original sentence of Coulson.

Aim of this Book is to focus on the properties of defects in semiconductors of the fourth group under a physico-chemical approach, capable to demonstrate whether the full acknowledgement of their chemical nature could account for several problems encountered in practice or would suggest further experimental or theoretical accomplishments.

Due to the poor information available concerning grey tin, analysis will be entirely devoted on diamond, silicon and germanium and on silicon- and germanium-carbides, which share the semiconducting properties of their elemental precursors.

It will be shown how difficult it can be, even today, after more than four decades of excellent research work, the fulfilment of self-consistency conditions, especially in the case of compound semiconductors (SiC in this book), but also in the apparently simplest cases of silicon and germanium.

1.1. Physical properties of group IV semiconductors

Group IV semiconductors include elemental and compound semiconductors, these last consisting of carbides.

C, Si, Ge and α-Sn (grey tin) share the common property to present a giant covalent structure in their diamond phases and to exhibit a rich number of allotropic phases, whose stability depends both on temperature and hydrostatic pressure, on which a vast literature exists [8].

Different from silicon, germanium and grey-tin, carbon presents the peculiarity of σ and π bonds formation, in addition to hybrid sp^2 and sp^3 bonds, which are not allowed for the other elements of the group. This makes the carbon chemistry different from that of silicon, as is well known, but could also influence some solid-state processes, as the bond reconstruction at surfaces, interfaces and grain boundaries of diamond [9] silicon and germanium.

Different from germanium and tin, the electronic structure of both carbon and silicon does not include the presence of d-electrons (see Table 1.1), but the empty d-orbitals in silicon are reasonably close to the highest occupied ground state orbital to allow their partial occupation by excitation or electron capture from strong electron donors, and contribute to bonding, with a valence shell expansion [10]. For this reason, silicon can form six-coordinated complexes, as $[SiF_6]^{2-}$, with its d-orbitals behaving as electron acceptors, like germanium and tin.

Different from carbon, silicon, germanium and tin are stable in two oxidation states (2+ and 4+), leading, for instance, to the following equilibria among oxides

$$SiO_2 + Si \rightleftharpoons 2SiO \qquad\qquad (1.1)$$

$$GeO_2 + Ge \rightleftharpoons 2GeO \qquad\qquad (1.2)$$

$$SnO_2 + Sn \rightleftharpoons 2SnO \qquad\qquad (1.3)$$

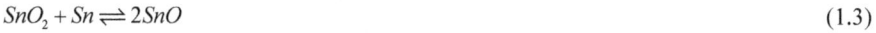

with the 2+ state being always less stable than the 4+ state. The situation is different with Pb, where the 2+ state dominates its chemistry.

Eventually, different from silicon, germanium and tin, which are synthetic elements, two natural allotropes of carbon are stable at room temperature and pressure, diamond and graphite, covalently bonded with bonds of sp^3 and sp^2 [5] character, while a recently discovered allotrope, the Q diamond, is structurally characterized by a mixture of sp^2 and sp^3 bonds and converts to diamond at room temperature [11].

Also, Si and Ge present several allotropes. As an example, using micro- and nano-indentation tests and annealing experiments, Kailer et al [12] showed the formation of at least 12 different high pressure silicon polymorphs. For both Si and Ge, the transformations sequence followed under applied pressure is first the transition to the β-tin structure with a coordination number (CN) =6, then to a simple hexagonal structure (CN=8) and finally to a hexagonal close-packed structure with (CN=12) [13].

Diamond is the cubic allotrope of carbon, and its structure (see Fig.1), with a tetrahedral coordination of the atoms in a cubic crystal lattice, is common to silicon and germanium in the entire range of temperature below their melting points, and to α -Sn at temperatures below the α-β phase transition at 286.35 K (see Table 1).

The first three elements of the fourth group of the periodic table present typical semiconducting character in the entire range of their existence as solid phases, while the forth element, tin, is metallic, with a close-packed (β) structure (density 7.265 g cm^{-3}), but presents a transition to the diamond (α) structure of grey tin (density 5.765 g cm^{-3} and to a (direct) semiconducting behaviour at temperatures lower than 13.2°C (see Fig.2a) [14].

As expected, the transition temperature from α- to β-Sn depends on the pressure [15], as shown in Fig.2b) [16].

---

[5] Graphite presents also van der Waals bonds.

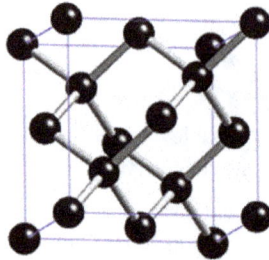

Fig. 1 Structure of diamond

a)

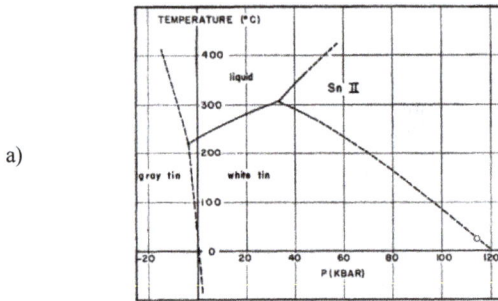

b)

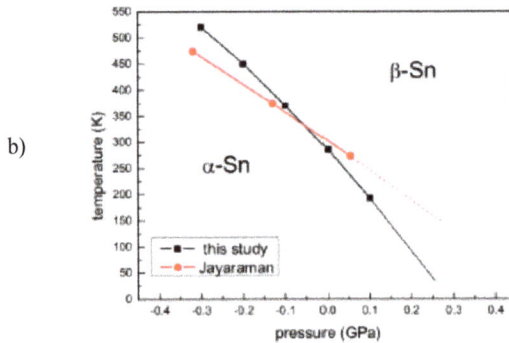

Fig.2 (a) Phase diagram of tin. After A. Jayaraman et al [14] Reproduced with permission from the American Physical Society, License number 4024281440475, License date Jan 08, 2017(b) Calculated pressure dependence of the α-and β -phases transition of tin. After S-H. Na, C-H. Park [16] Creative Commons Attribution 4.0 International (CC BY).

The α- β transition of tin occurs after a long induction period, but it has been shown [17] that a preliminary reactor irradiation, with a neutron flux of 10^{18} nvt^{-1} at the liquid nitrogen temperature, drastically reduces the induction period, measured in the temperature range -60 to 20 °C, indicating that the defects introduced into white (β) tin by reactor irradiation served as nuclei or possible embryos for the phase transformation. Although it cannot be excluded the influence of the reactor irradiation strain around displaced atoms on an easier transformation, still the hypothesis that defects are the predominant cause fits well with later results on the issue, which do evidence the presence of lattice defects induced by the transformation [18]. The question concerning the thermodynamic stability of the α-phase with respect to the β-one has been, eventually, discussed considering that the Sn-3d orbital states are located far below the Fermi level for each of the two phases, but in the semiconducting α-phase the s and the p orbitals are split around the Fermi level. This condition favors a strong hybridization of the s and the p orbitals in α-Sn, raises the energy level of the anti-bonding states at the conduction band and lowers the energy level of the bonding states at the valence band [16]. It could then be supposed that the lowering of the electron-occupied bonding states could contribute to thermodynamic stabilization [16, 19-21] of the α-phase respect to the β-one.

Lead, the fifth element of the group, is instead, metallic in the entire range of stability of his solid phase, and the transition from a metallic- to a semiconducting-character when passing from lead to α- Sn depends on the fact that the bond length of grey tin is small enough to allow the formation of hybrid sp^3-bonds with the consequent opening of a (small) band gap [22].

The strong covalent bond character does account not only for the semiconducting properties of these elements in their diamond structure, but also for a peculiar feature of diamond, silicon and germanium in their liquid phases, which has been the subject of extensive experimental and theoretical research, recently.

As an example, silicon presents at the melting temperatures a transition to a pseudo-metallic liquid phase that could be considered to consist of a mixture of a metallic- and covalent-liquid phase [23-24], with a persistence of covalent bonding in the liquid phase.

The case of germanium is slightly different, since the liquid resembles a mixture of two phases, of which one consists of the regular metallic liquid and the second is a vacancy-rich phase according Davidovich [25]. According to Jank and Hafner [26], furthermore, the calculated electronic structure of liquid germanium is far from any of the semiconducting and metallic crystalline phases of germanium. Compared to the metallic high-pressure phases one observes, eventually, in liquid Ge higher interatomic distances which lead to a much smaller orbital overlap and to a break of the plane-wave hybridization typical of metals.

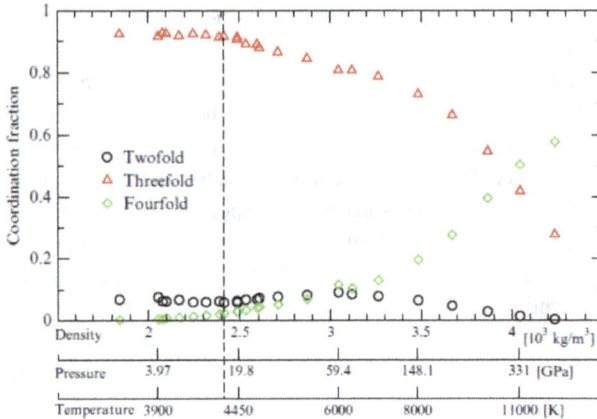

Fig.3 Coordination fraction of the liquid along the melting line of graphite (the dashed line marks the liquid/graphite/diamond triple point) and diamond.   After L. M. Ghiringhelli and E. J. Meijer [28]. Reprinted with permission from Springer, License nr. 3979460083352, Oct.31, 2016

The structure of the liquid carbon near the freezing point of graphite and diamond has been eventually modelled with Monte Carlo simulations and DF-MC[6] calculations using the Car-Parrinello method, showing (see Fig. 3) that contrary to the generally assumed picture [27] for which diamond melts into a four-coordinated liquid, both graphite (left to the crossed line) and diamond melt into a three-coordinated liquid, which is gradually substituted by a four-coordinated liquid with the increase of the pressure and density [28]. The mass amount of the two-fold coordinated liquid remains instead negligible.

---

[6] DF-MC means Density Functional-Monte Carlo.

1.2   Chemistry and thermodynamics of group IV elements

Of the group IV elements, only diamond and graphite could be found in nature as almost pure elemental materials: all other elements are found in compounds and, therefore, elemental semiconductors are synthetic materials. The trivial reason of it is that the main natural oxidation products of carbon are gaseous phases (CO and CO_2), while the other group IV elements form stable compounds by oxidation that are found as such in nature.

One can see in Table 1 that lattice spacing, bond length and density [7] of group IV semiconductors increase systematically from diamond to Sn, with a corresponding decrease of the bond- and cohesive-energies (see Table 2).

Table 1. Physical properties of group IV semiconductors [20].

Element	Atomic number	Electronic structure	Structure	Lattice spacing (nm)	Density (g cm^{-3})	Pauling electrone gativity
C	6	$1s^2 2s^2 2p^2$	Diamond	0.35668	3.51	2.55
Si	14	$1s^2 2s^2 2p^6 3s^2 3p^2$	Diamond	0.543071	2.328	1.90
Ge	32	$1s^2 2s^2 2p^6 3s^2 3p^6 3d^{10} 4s^2 4p^2$	Diamond	0.565791	5.323	2.01
Sn	50	$1s^2 2s^2 2p^6 3s^2 3p^6 4s^2 3d^{10} 4p^6 5s^2 4d^{10} 5p^2$	Diamond (below 286.35 K)	0.6489	5.765 (at 20°C)	1.96

Table.2. Thermodynamic properties of group IV elements.

	Bond (dissociation) energies (kJ/mol)	Band gap energies[3] (eV)	Cohesive energies (eV)	Bond length (pm)	Melting temperature (K)
Diamond	346	5.6	14.7	154	3800
Silicon	222	1.107	7.75	233	1687
Germanium	188	0.67	6.52	244	1211.2
Tin	146	0.079	5.5	280 (302)	505 (β-Sn)
Lead	339	0		350	327.5

It can also be seen in Table 2 that the band gap energies decrease almost linearly from diamond to grey silicon with the decrease of the cohesive energy and of the bond energies.

---

[7] except for diamond.

As can be seen in Table 2 and in Fig. 4, also the melting and boiling temperatures of group IV elements decrease from diamond to lead.

Solid solubility is drastically limited, as expected, in the case of C-Si, C-Ge, Si-Sn and Ge-Sn systems [29-30] (see in Fig. 5 the Si-Sn and Ge-Sn phase diagrams), due to significant differences in the bond lengths of the elements involved.

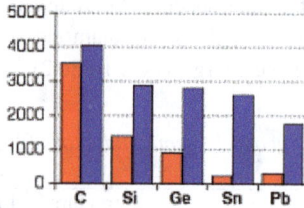

Fig.4 Melting (in red) and boiling points of group IV elements

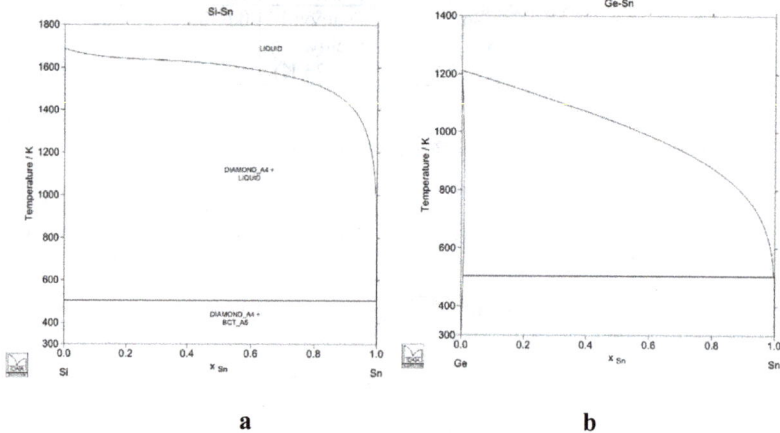

a   b

Fig.5 (a) Calculated phase diagram of the Si-Sn System (b) Calculated diagram of the Ge-Sn system after MTDATA-Phase Diagram software from the National Physical Laboratory.

A drastic consequence of the limited solubility of Sn in Ge is the impossibility to prepare Ge-Sn alloys with 10 % of Sn, a composition at which the onset of direct gap conditions is predicted and important optoelectronic applications[8]are foreseen.

Common structures, close lattice spacing and similar electronegativities favor, instead, in the case of silicon- germanium solutions, the onset of complete solid solubility and of an almost ideal thermodynamic behaviour.

As it is well known, the thermodynamic behaviour of a semiconductor phase consisting of a multinary solid solution of a main component (C, Si, Ge, Sn) and of several impurities could vary from

i)   that of an ideal binary solution, for which interactions between the solution components A and B are negligible and, consequently, the enthalpy of solution ΔH_{sol} is zero or negligibly small, and the Gibbs free energy of solution corresponds to the configurational entropy ΔS_{conf} contribution

$$\Delta G_{sol} = -RT(x_A \ln x_A + x_B \ln x_B) = -T\Delta S_{conf} \qquad (1.4)$$

ii)   to that of a non-ideal multinary solution, with large values of the enthalpy of solution, that is function of an interaction parameters Ω accounting for chemical and mechanical interactions

An intermediate case could be foreseen and experimentally observed, that of binary regular solutions, for which a simple relationship exists between the solution composition and the enthalpy of solution (or mixing enthalpy), which is given by the product

$$\Delta H_{sol} = x_A x_B \Omega \qquad (1.5)$$

with $\Omega = \left(H_{AB} - \dfrac{H_{AA} + H_{BB}}{2} \right)$, where H_{AB} is the binding energy of the AB species and

H_{AA} and H_{BB} are the binding energies of the A and B species, respectively.

An interaction parameter Ω close to zero (Ω= 0.01 eV) is experimentally found for Si-Ge solutions, that minimizes the contribution of mixing enthalpy $x_{Si}x_{Ge}\Omega$ (~ 1.3 kJ/mol [31] to the Gibbs free energy of mixing

$$\Delta G_{sol}(x,T) = x_{Si}x_{Ge}\Omega - T\Delta S_{conf} \qquad (1.6)$$

---

[8] Epitaxial deposition on Ge substrates of metastable, Sn-rich solid solutions has been shown a viable solution.

where x_{Si} and x_{Ge} are the molar fractions of Si and Ge in solution, and leads to a quasi-ideal behaviour

$$\Delta G_{mix}(x,T) = -RT\left[x_{Si}lnx_{Si} + x_{Ge}lnx_{Ge}\right] = -T\Delta S_{conf} \qquad (1.7)$$

with a purely entropic character of the Gibbs free energy of mixing.

The distribution of silicon and germanium atoms in the solid solution is expected, therefore, to be almost completely random.

It is, however, interesting to observe that:

- the composition dependence of the lattice parameters of Si-Ge alloys significantly deviates from the Vegard law[9] [8], that would predict for the lattice spacing $d_{Si_xGe_{(1-x)}}$ of Si_xGe_{1-x} alloys a linear composition dependence

$$d_{Si_xGe_{(1-x)}} = xa^o_{Si} + (1-x)a^o_{Ge} \qquad (1.8)$$

  where a^o_{Si} is the lattice constant of pure silicon and a^o_{Ge} is the lattice constant of pure germanium, and

- the distance between neighbors, that correspond to specific bond lengths, vary only little with the composition of the alloy [32] (see Fig. 6).

The last feature is an additional confirmation of the almost ideal thermodynamic behaviour of Si-Ge solutions, as the limited bond length dependence on composition would imply limited Si-Ge interaction.

---

[9] The Vegard's law is a purely phenomenological law, and as such, severely violated in a number of cases [8].

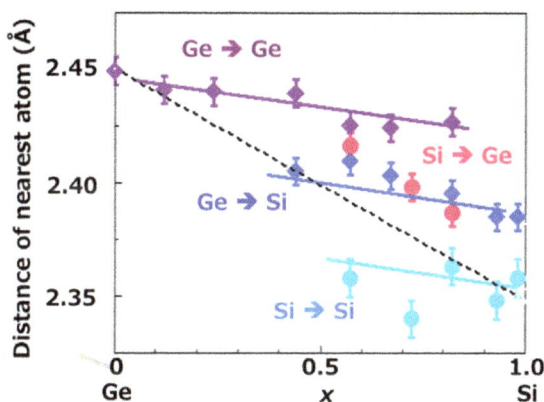

Fig.6 Composition dependence (in molar fraction) of covalent bond lengths in silicon, germanium and their alloys. With kind permission from the H.Yonenaga, Tohoku University, Sendai, Japan.

The monotonic variation of the lattice constant of the alloy is, therefore, accomplished by a prominent contribution of specific bonds (Si-Si, Si-Ge and Ge-Ge) to the average lattice spacing, in specific composition ranges, as could be easily inferred again from Fig. 6.

1.3   Chemistry and thermodynamics of group IV carbides

Alloys and carbides of group IV elements are today granted of immense potential for future technological applications of C-based electronic and photonic devices [33], using single crystal carbides as such or as homoepitaxial and heteroepitaxial films on SiC and silicon single crystal substrates.

Ordered compounds and C-based alloys of Si, Ge and Sn present, or theory foresees to present, in fact, semiconductor properties like their elemental precursors[10], together with specific, individual thermodynamic and structural properties.

However, while SiC is the intermediate, stable (ordered) compound in the Si-C phase diagram, see Fig.7, and SiC, in fact, segregates from a C-saturated liquid silicon, no evidence of compound formation is found in the case of germanium, up to temperatures

---

[10] For SiC the semiconducting properties are already well known and industrially well exploited..

of 3170 °C, as demonstrated early in 1959 by Scace and Slack [34], who showed, also, that the solubility of carbon at the melting temperature of silicon is 3.0 ±0.3x 10^{18} at cm^{-3}, while is only 10^8 at cm^{-3} at the melting point of germanium.

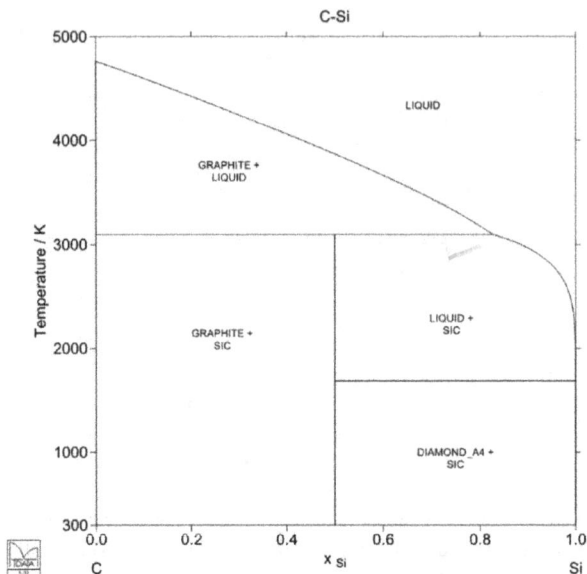

Fig.7 Calculated diagram of the C-Si system after MTDATA-Phase Diagram software from the National Physical Laboratory.

Later work [35], addressed at the determination of the carbon solubility in Ge, and carried out using a nuclear technique and ^{14}C as the tracer, showed that the carbon solubility is much larger than that determined in [34], since it amounts to about 10^{13} at cm^{-3}, still orders of magnitude smaller than in silicon.

Eventually, Both and Voss [36] confirmed that a form of GeC analogous to SiC is not known to exists as a naturally occurring crystalline phase, as is also the case of SnC.

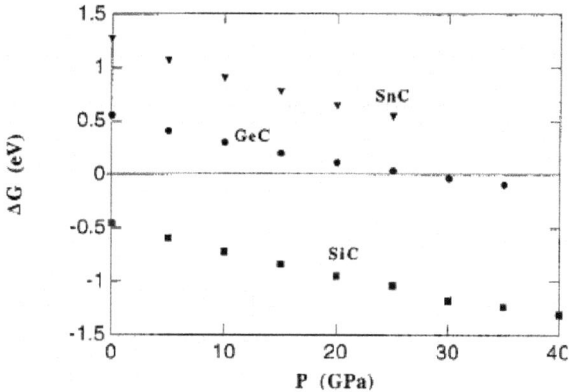

Fig.8 Calculated free energy of formation of SiC, GeC and SnC. After R. Pandey et al [33]. Reproduced with permission from AIP Publishing LLC, License number 4020310055207, License date Jan 01, 2017.

The low or negligible thermodynamical stability of the GeC and SnC phases, as ordered compounds of Ge and C and of Sn and C, has been theoretically proven by Pandey *et al* [33] using DFT calculations. As can be seen in Fig. 8, GeC begins to be thermodynamically stable only at pressures higher than 30 GPa, while SnC is thermodynamic instable up to the highest pressures.

The structural and electronic properties of SiC and GeC have also been calculated by C.A. Madu [51] using the Density Functional Theory (DFT). The good fit of the calculated values of the lattice constant of the cubic structure of SiC (4.362 Å vs the experimental value of 4.36 Å) validates the calculated value (4.568 Å) of the lattice constant of the hypothetically stable GeC phase.

Eventually, it has been also calculated that GeC should be stable in the zincblende (E_g = 1.76 eV) and wurtzite (E_g = 2.5 eV) structures, behaving like an indirect gap semiconductors, and presenting a high degree of covalence (80-90 %) [48-50], while $Ge_{1-x}C_x$ alloys should exhibit direct gap properties, with superior optoelectronic properties. Details on this issue are reported at the end of this section.

The limited solubility of carbon in Ge adds a limit to germanium application in microelectronics, as it virtually impedes the CVD growth of $Ge_{1-x}C_x$ alloys, which could present interesting structural properties, since C-addition to Ge would shrink the lattice constant of Ge close to that of Si, potentially allowing the direct heteroepitaxial deposition of Ge on Si without making use of buffer layers.

Despite the forecast of superior technological promises of GeC and of $Ge_{1-x}C_x$ alloys, today only SiC carbide deserves the full interest for electronic and optoelectronic applications, and found very important uses for high power diodes and high frequency devices.

As it is well known, SiC belongs to a class of materials dominated by polytypism, a kind of polymorphism which depends on the different order of stacking layers of Si-C polyhedra in the SiC lattice [8]. Three types of layers exist, labeled A, B, C, which allow easy silicon and carbon packing, and the difference between the different polytypes depends on the stacking sequence adopted.

Two main structures are known for SiC, the cubic (zinc-blende) β-SiC and the hexagonal α-SiC (see Fig. 9), together with several (more than 200) other polytypes.

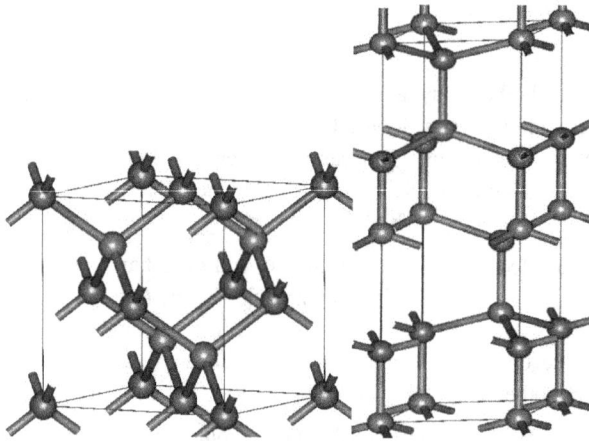

Fig. 9 Structure of cubic β-SiC (left) and of hexagonal α-SiC (right).

The same structures are predicted for GeC in the range of pressures enabling its stability.

The preparation of single crystal SiC [8] starts with the synthesis of raw SiC, prepared by the carbothermal reduction of the oxide, via the Acheson process, which could be conveniently described by the following equation

$$SiO_2 + 3C \rightleftharpoons SiC + 2CO \tag{1.9}$$

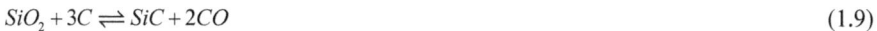

but is kinetically dominated by reactions involving silicon monoxide in a true vapor-phase process

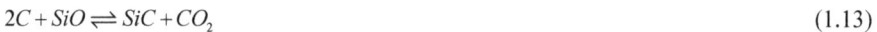

$$C + SiO_2 \rightleftharpoons CO + SiO \tag{1.10}$$

$$SiO_2 + CO \rightleftharpoons SiO + CO_2 \tag{1.11}$$

$$C + CO_2 \rightleftharpoons 2CO \tag{1.12}$$

$$2C + SiO \rightleftharpoons SiC + CO_2 \tag{1.13}$$

The material obtained from this process is used for the growth of single crystal boules via the Lely process [8], which occurs *via* the sublimation of SiC powder under argon at temperatures ranging between 2200 and 2450 °C [42-44].

The carbothermal synthesis of GeC from GeO_2 is expected, at least, to be less effective than that of SiC from SiO_2, due to its intrinsic thermodynamic instability and the poor thermodynamic stability of GeO at temperatures below 600 °C [45]. Still a material formally consisting of 99 to 99.999 % GeC is manufactured and sold by an USA Company.

Despite the widespread application of SiC single crystals in microelectronics, thin film Si-C alloys present great prospective interest in view of potential cost reductions of the SiC chain and better integration of SiC in microelectronic applications, as is the case of their use as substrates for Micro-Electro-Mechanical Systems (MEMS) applications [39]. Additional advantages would result from their piezo-resistive properties, that favor their use in pressure sensors and accelerometers [38] applications.

Thin film SiC have been prepared with several techniques [40-41], including Plasma enhanced CVD (PECVD), Low Pressure CVD (LPCVD), Molecular Beam (MB) epitaxy and Magnetron sputtering.

SiC thin films deposited at ambient temperature are generally amorphous, and their structure/composition depends on adopted growth parameters.

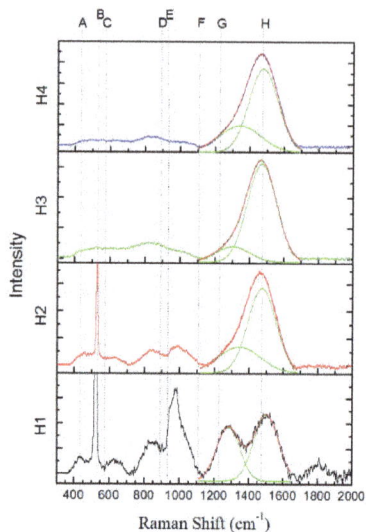

Fig.10 Raman spectra of four SiC samples prepared with the HiPIMS technique. After G. Leal et al [41] On-line version ISSN 1980-5373.

Table 3. List of the Raman lines and of the corresponding bonds of the SiC samples of Fig.10 [41].

Line	Exp.wavelength (cm⁻¹)	Literature wavelength	Bonds
A	435	470	Si-Si
B	528	521	Si-Si
C-D	577-906	550-1000	SiC
E-F	931-1103	750-1100	Si-Si
G	1288	1300	C-C (D band)
H	1480	1500	C-C (G band)

18

Fig. 10 displays, as a non-exhaustive example, the evolution of the Raman spectra[11] of amorphous SiC films deposited on silicon substrates at room temperature, using the High-Power Impulse Magnetron Sputtering (HiPIMS) technique at increasing power. Table 3 lists the positions of Raman peaks and the corresponding bonds [41].

One can see that with the increase of the applied power (and the increase of the film thickness) the Raman lines of the crystalline silicon (B, E, F) obviously disappear, while the intensity of the SiC lines (C-D) decreases with the increase of those of C-C bonds. Low power conditions are, therefore mandatory for good thin film preparation.

Polycrystalline and monocrystalline silicon carbide thin films were also prepared using Cyclotron Resonance Enhanced Plasma (CREP) [46], and the lowest temperature range at which β- SiC has been deposited lies around 690-900 K.

As expected, the CVD preparation of Ge_xC_y alloys deserved, instead, null or moderate success.

As an example, Both and Voss [47] used GeH_4 and acetylene (C_2H_2) as Ge- and C-sources and fused quartz substrates at deposition temperatures between 475 and 550 °C, and showed that the material prepared by CVD is a heterogeneous mixture of Ge clusters and of a Ge_xC_y phase of unknown composition.

A moderate success was instead obtained by Herrold and Dalai [37], who used a plasma process with GeH_4 and CH_4 as source gases and a substrate temperature of 350 °C, using glass, steel and silicon substrates, to CVD grow C_xGe_{1-x} alloys.

The best material obtained was a microcrystalline solid solution of carbon in Ge, with a maximum C-content in the alloy of 1.5 at%, as determined from Vegard's law calculation of lattice contraction as a function of the substitutional carbon concentration. The best samples were obtained using silicon as the substrate.

---

[11] Raman spectroscopy is systematically applied in SiC characterization

Chapter 2

2.   Defects in group IV semiconductors

2.1   Introduction

Like all common thermodynamic systems, group IV elemental semiconductors as single crystals, polycrystalline samples and epitaxial layers, should present significant deviations from their ideal structural perfection and chemical identity, since disorder would favor their thermodynamic stability associated to an increase of configurational (and vibrational) entropy.

Local deviations from structural and chemical perfection of ionic solids, metals and semiconductors are traditionally called defects, which could present, formally, either point-like or extended features.

Point defects (vacancies, self-interstitials, substitutional and interstitial impurities) are (ideally) 0D defects, whose effective presence and concentration in a crystalline solid would depend on the history of the sample, i.e. on the growth procedures and on conventional/unconventional thermal processes[12] adopted for the preparation of usable samples[13]. In the case of diamond, the only non-synthetic semiconductor of the IV group[14], the defect nature and concentration should depend as well on its synthesis in natural geothermal conditions.

At ambient temperature, native point defects (vacancies and self-interstitials) are *frozen* in a non-equilibrium state, whose actual properties depend on the high temperature and pressure conditions at which defects were generated under equilibrium or close to equilibrium conditions, on the rate of the cooling process and on the nature and chemical quality of the medium in which cooling occurred.

The presence of impurities, in fact, induces the occurring of defect-impurity interaction processes of various types, on which a vast literature exists.

When the defect properties of a synthetic or natural semiconductor sample are investigated after having established novel equilibrium conditions by a suitable thermal treatment, to follow their temperature dependence, one should consider that the measured

---

[12] Thermal annealing, diffusion and ion implantation process.

[13] by suitable synthesis routes and growth techniques.

[14] We ignore here the properties of synthetic diamond.

equilibrium properties account for all chemical (and mechanical) interactions occurred among all system's components (defects and impurities) during the thermal process.

Therefore, the experimental defect properties, as an example, the defect formation energy, might differ from those evaluated with *ab initio* or Molecular Dynamics (MD) calculations, that refer to an ideal crystal (here an impurity-free semiconductor) where chemical interactions are hypothetically absent. These last properties could be, however, used as reference values.

Extended defects are 1D, 2D or 3D defects (dislocations, internal and external surfaces, precipitates) are non-equilibrium defects which originate from mechanical stresses and chemical processes occurring during post-growth or device manufacturing processes.

Their very identity and properties might also depend on chemical interactions with impurities, which could arise from unwanted or deliberately induced chemical contamination of the sample during its preparation or synthesis.

Defect studies in semiconductors have been and still are an exciting example of theoretical and experimental excellence, with relevant consequences on semiconductor science and technology.

Defects in group IV semiconductors have been deeply studied in the last sixty years, starting from silicon and germanium, but with an increasing emphasis to diamond in the last decades, with thousands of papers and hundreds of excellent textbooks and Symposium Proceedings, devoting their major emphasis on the microscopic, spectroscopic and electronic properties of defects in diamond, silicon, germanium, Si-Ge alloys and silicon carbide [52-60].

The next sections will address the analysis of the physico-chemical aspects of point defects in silicon, germanium, diamond and silicon carbide, since a knowledge of defects in grey tin is practically absent, mostly due to the difficulty to grow it in single crystal structures of appreciable size [61].

2.2   Physico-chemical properties of point defects in the diamond lattice: experimental results and theoretical modelling

2.2.1   Theoretical and experimental evidences

Empty reticular sites (vacancies) and atoms in interstitial positions (self-interstitials) [15] are the native point defects in group IV semiconductors that, in analogy with similar defects present in ionic solids, metals and other elemental and compound semiconductors, when generated under proper thermal excitation, should behave as true quasi-chemical species in thermodynamic equilibrium.

Two kinds of geometrical configurations could be theoretically supposed and theoretically modelled for vacancies embedded in a diamond lattice of which V_L is the stable lattice defect (see Fig. 11a), and V_B, called split-vacancy, is the transition state of a vacancy in the due of a jump from one lattice position to a neighbor one [16] [62]. Several geometrical configurations could be, instead, imagined and modelled for self-interstitials (see Fig. 11 b, c, and d), as the split -one (I_{100}), the tetrahedral (I_T) and the hexagonal one I_H, of which the I_{100} configuration is the most stable [63].

Vacancies and self-interstitials as individual defects generated in thermodynamic equilibrium conditions are, however, elusive particles, since their presence is difficult to be experimentally detected due to their almost negligible concentrations.

A deep insight on their properties (structure, energy of formation and spectroscopic signatures) arises, instead, from theoretical investigations based on *ab initio*, Molecular Dynamics (MD) and Density Functional Theory (DFT) calculations, which started in the late eighties of the last century [63-67] and gained a renewed emphasis in more recent years [62-63, 68-72].

The presence of defects has been, instead, experimentally demonstrated in samples submitted to high energy -X-ray, γ-ray, and -electron irradiation or ion- implantation for scientific experiments or technological[17] treatments, when the energy of the irradiating species or photons is taken above the displacement threshold energy of the atoms in the lattice, whose value depends on the nature of the material, that in diamond ranges between 37.5 and 47.6 KeV (3521.7 to 4592.7 MJ/mol), depending on the crystal orientation [73].

---

[15] Impurity atoms could also sit in interstitials positions and are as well interstitial defects, but non-necessarily equilibrium defects.

[16] Atomic (and defects) motion in solids does occur with jumps from one lattice position to the nearest neighbour one, if tunnelling processes are ineffective.

[17] Ion implantation is a routine technological process for semiconductor doping.

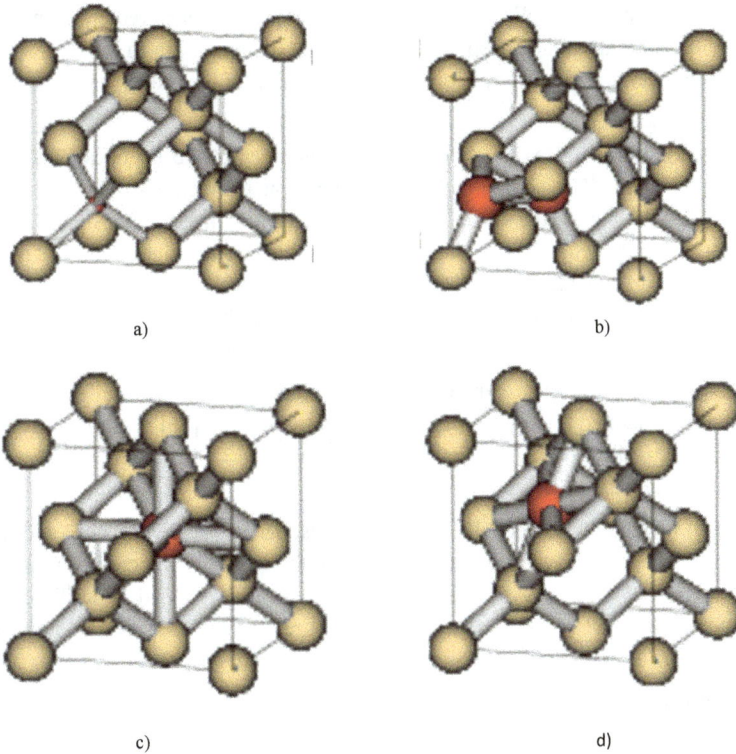

Fig.11 Structure of a vacancy (a) and of $I_{<100>}$ (b) I_T (c) and I_H (d) interstitials in the diamond lattice. After S.A. Centoni et al [68]. Reproduced with permission from the American Physical Society, License nr. 3995400587074 Nov.24, 2016.

In these conditions a supersaturated amount of vacancies and self-interstitials is generated (together with defect complexes and radiation damage centres), whose presence could be detected by spectroscopic measurements that provide their signatures and further insights for their structural characterization.

Low temperature irradiation and measurements could be needed to preserve the defect integrity, if thermal annealing after irradiation would favor defect-defect and defect-impurities interaction, as it is worthwhile for Si, where vacancies become mobile at T> 200 K, leading to their recombination and point defect complexes formation [55,74-75], as will be shown later in details.

Due to the presence of Dangling Bonds (DBs) and associated unpaired electrons and spin states (which could introduce deep levels in the gap) Electron Spin Resonance (ESR) had at the very beginning [6] a key role on the determination of the properties of vacancies in silicon.

This is not the case of Ge, for which the high number of Ge isotopes [76], the high nuclear spin of the ^{73}Ge isotope and strong spin-orbit interactions [77] were supposed factors capable to impede the vacancy detection. Recent numerical investigations carried out using hybrid functional calculations [70] show, instead, that DBs in Ge do not generate levels in the band gap. DBs behave as negatively charged acceptors, whose energy level is below the valence band maximum. This better explains the absence of ESR activity of DBs in germanium.

Positron Annihilation Spectroscopy (PAS) is demonstrated to be well suited for the study of vacancies, generated by low temperature proton irradiation [76], as shown for example by Meher [56], while Deep Level Transient Spectroscopy (DLTS) is of general use in the case of electrically active defects in Si and Ge.

Eventually, Optically Detected Magnetic Resonance (ODMR) in addition to Photoluminescence (PL) has been shown of use in the case of semiconductors presenting optically- active defect centres, as is the case of diamond [78-79] and silicon carbide (SiC), and Raman spectroscopy is demonstrated of relevant interest for the detection and characterization of neutral vacancies in irradiated diamond [80].

The properties of isolated self-interstitials remain, instead, elusive, due to the practical absence of bond exchanges with the host matrix, *i.e.* of chemical activity, leading to a very limited number of experimental evidences.

As an example, due to the presence of two dangling bonds at its core, the I_{100} self-interstitial should be ESR active in its neutral state and therefore spectroscopically detectable in group IV semiconductors. This hypothesis has been, in fact, demonstrated valid in the case in diamond [81], where the R2 ESR centre is attributed to self-interstitials, but not for Si and Ge, where the concentration of neutral I_{100} self-interstitials is shown to be undetectably small [60]. Instead, an ESR centre associated to the caged T_d interstitial is observed in intrinsic-or p-type silicon after irradiation with protons [69].

Their detection has been found, also, indirectly possible by stimulation of the formation of self-interstitial- impurity complexes [6], but the matter remains very intriguing, due to severe difficulties arising when looking to their precise chemical attribution, different from the case of complexes involving vacancies centres, on which we will deal in details in Section 2.4.

As an example, an ESR active centre in diamond, associated to a photoluminescence emission at 503.5 nm is attributed to a complex involving a single self-interstitial and an unknown impurity, according Steeds *et al* [82]. Several other self-interstitials complexes with unknown impurities have been further detected by PL spectroscopy [82-83] in X-ray irradiated diamond above the carbon displacement threshold, showing, also, that self-interstitials in diamond are mobile with an activation energy of 1.3 eV at temperatures below 50 K and form several impurity-interstitial complexes that decompose above 400 K. It has also been experimentally shown that the self-interstitials released from the decomposition of these complexes interact with the vacancy-nitrogen (NV) centres causing their decomposition.

The presence of self-interstitials, together with vacancies, is however, indirectly evident in silicon and germanium at device process temperatures, since dopant- and impurity-diffusion processes could not be understood and modelled unless supposing that the diffusion process itself is mediated by vacancies and self-interstitials [84-85], in view of the negligible contribution of Direct Exchange processes, as we will show in details in the next Section.

Eventually, self-interstitials and vacancies behave as equilibrium or quasi-equilibrium species in the high temperature growth and post growth processes of semiconductors, where they determine, as an example, the segregation of oxygen and carbon in silicon and the features of the microscopic defects consisting in aggregates of vacancies and self-interstitials, which evolve during the growth of single crystal Si^{18}, as shown by Voronkov in his classical paper [86].

2.2.2 Structure of point defects

Different from point defects in metals and ionic solids, the geometrical, point-like features of vacancies and self-interstitials in the covalent lattice of group IV semiconductors shown in Fig. 11, but of all other semiconductors, are purely formal, and their localized or extended features have been debated for years.

Among others, Seeger and Chick [87] introduced the concept of spatial extension for lattice defects in silicon and germanium, that was severely questioned by Fahey *et al* [88], arguing that if a defect loses its point-like features, its migration properties are not amenable in a simple way.

---

[18] The growth process of single crystal ingots starts at the melting temperature of the semiconductor and defect generation processes do occur during the ingot cooling so far the temperature is sufficiently high to permit defect motion and defect/impurity interaction.

The spatial extension of lattice defects is actually supported by the use of modern computational methods, as demonstrated by Cowern et al [89-90], who showed that while at low temperatures the vacancies and self-interstitial behave like true point-defects, at high temperatures an extended quasi-amorphous region (morph) consisting of $N\pm1$ lattice atoms, which incorporates either a self-interstitial or a vacancy, is thermodynamically stable. As an example, a vacancy-type morph could consist of N-1 atoms unit in a lattice island of N atoms, while an interstitial-type morph would consist of a N+1 atoms unit in a lattice island of N atoms. It is also supposed that the morphs have regular coordination with the host lattice, but that their inner structure presents the coordination typical of an amorphous phase.

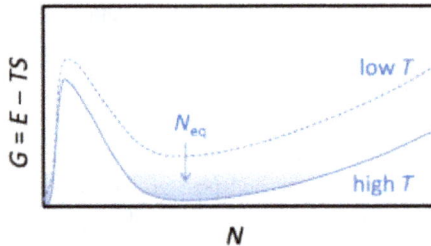

Fig. 12 Schematic diagram of the Gibbs free energy of a compact self-interstitial (dotted curve) and of a morph (continuous curve) as a function of N. N ranges between 1 and 2 for the compact self-interstitials and between 1 and 40 for the morph. After N.E.B. Cowern et al [89]. Reproduced with permission from the American Physical Society, Delivery Date Nov. 05,2016 License number 3982611322370.

A schematic diagram displaying the calculated Gibbs free energy of a compact self-interstitial (upper curve) and of a morph is reported in Fig. 12. One can observe that the minimum of the Free Energy curve is split from N=2 in the case of a compact self-interstitial toward 20 for the case of a self-interstitial embedded in a morph.

According to Cowern et al [89-90], migration of these complex structures depends on peripheral rebonding, which leads to rough estimates of self-diffusion activation energies of the order of 6 eV for both Si and Ge.

In the hypothesis of morph existence, since the formation- and migration-energies depend on the number of atoms in the morph, it could be understood why the high temperature diffusivity of vacancies and self-interstitials are so close in the case of silicon, as we will see in Section 3.

Independently of morphs existence, however, the extended nature of vacancies in diamond-type of structures could be envisaged considering that their formation involves covalent bond breaking, with further bond pairing or reconstruction, empty volume formation, lattice relaxation and bond distortion, as illustrated by Watkins already four decades ago [6]. This should lead to a final configuration that is necessarily extended in the space and ideally consists of a missing atom in the centre of a regular tetrahedron, as seen in Fig. 13.

The discussion of the structure of vacancies in a covalently bounded lattice is conveniently carried out using a molecular-like model, originally used by Coulson and Kearsley early in 1957 [5] *assuming that in a non-polar crystal the vacancy electrons do not "spill over" from the vacancy in the rest of the crystal and that the levels of the vacancy electrons can be considered without making allowance of exchange or polarization by other electron clouds[19]*. Coulson and Kearsley assumed, also, *that in analogy with the situation in conventional molecules, we do not expect that the adjacent bonds will be greatly affected.*

Nevertheless, the situation of a vacancy in a covalent crystal is rather different from that of an isolated molecule.

Ideally, when an atom is removed from a lattice position of a diamond structure and a vacancy is formed in the centre of a regular tetrahedron, it leaves four dangling bonds (DBs) which behave like chemically active radicals and make such defect very reactive.

The system could reach its relative thermodynamic stability by bond reconstruction or by bonding with impurities or other defects, as is the case of formation of the vacancy-oxygen centres or of a di-vacancy.

The different geometrical configurations that could be obtained by partial or total saturation of DBs by impurities or by other defects will be discussed in Section 2.2.

---

[19] original sentence in the Coulson paper.

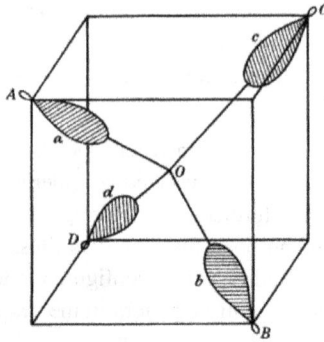

Fig. 13 Defect molecule for an isolated vacancy. After Coulson) [5]. Reproduced with permission of the Royal Society, License number 4039270233956, License data Jan 31, 2017.

In the absence of external interaction sources, the system reaches its equilibrium configuration (see Fig. 14) with a DFT-calculated inward relaxation of the four atoms neighbor to the vacancy toward the vacant site from an initial distance of 235 pm to 222 pm.

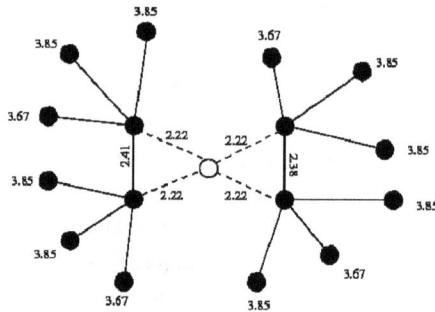

Fig.14 Schematic picture of a silicon vacancy after bond reconstruction. The different lengths of the reconstructed bonds are indicated. The equilibrium Si-Si bond length is 235 pm. Reproduced with kind permission from prof. Stewart Clark, University of Durham (UK): after S. Clark [91].

Relaxation is followed by dangling bonds saturation and pairwise bond reconstruction to form two non- interacting pairs covalently bonded [91]. After bond reconstruction, also the second neighbor distance changes with an inward relaxation from 385 pm to 367 pm. Incidentally, this configuration has an *ab initio* calculated formation energy of 3.38 eV,

≈ 0.5 eV less than the total energy, DFT-calculated value of the neutral vacancy (5 eV) by Car *et al* [63], evaluated without considering short range relaxation phenomena, with uncertainty ranging from 0.5 to 1 eV.

This process occurs with the breaking of the original T_d symmetry toward a tetragonal D_{2d} distortion, due to the Jahn- Teller effect [70-71, 92-94].

The extended nature of the vacancy leads to pressure dependent energies of formation, as is shown in Fig, 15, where one can observe the calculated effect of hydrostatic pressure on the energy of formation of vacancies in the T_d and D_{2d} symmetry, as well of split vacancies.

It is apparent that the Jahn-Teller distortion reduces the enthalpy of formation and results in a volume contraction of about $0.5\,\Omega$ (bottom scale), where Ω is the atomic volume of silicon. As expected, relaxed vacancies with D_{2d} symmetry are less sensible to pressure effects.

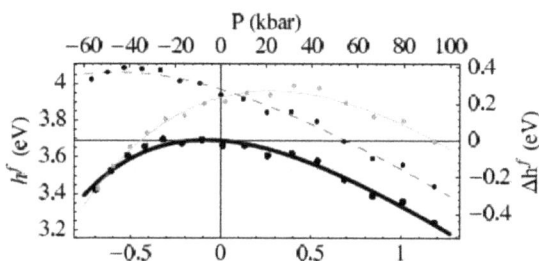

Fig. 15 Calculated effect of hydrostatic pressure on the formation enthalpy of vacancies (Full heavy line: relaxed vacancies with D_{2d} symmetry, dashed line: V^oB (split vacancies) and grey line: vacancies with T_d symmetry. After S.A.Centoni et al [68] open access Journal / Creative Commons Attribution license (3.0 Unported or 4.0 International).

Also individual self-interstitials may be extended, as is the case of the split I_{100} one, and of the caged-interstitial, that is a relaxed, metastable structure of the I_{100} selfinterstitial [95].

The extended nature of defects in semiconductors should have a non-negligible effect on the activation energy of diffusion, since the local 3D environment of a defect should be transferred during a diffusion step. Therefore, the activation energy values include entropic and enthalpic terms associated to defect localization/delocalization and bond reconstruction terms. This issue will also be considered in the next section, dealing with the rate of recombination of vacancies and self-interstitials, which should be discussed with the inclusion in the rate constant of enthalpy and entropy terms associated to the defect re-localization.

2.2.3 Generation of equilibrium point defects, thermodynamics and kinetics

The fact that "point" defects in semiconductor crystals are extended in space, does not exclude that they could behave as equilibrium species, that their formation was the result of a chemical reaction and that their concentration, in the absence of an excess of photo-induced- or irradiation- generated defects[20], must obey to the mass action law, provided the temperature is sufficiently high to suppress, or make negligible, possible kinetic hindrances.

On that basis, the generation of isolated, equilibrium point defects, vacancies and self-interstitials, might be formally represented by one homogeneous (bulk), and two heterogeneous chemical reactions [88].

The homogeneous one, also known as the Frenkel process, foresees the formation of a couple of isolated vacancy (V) and self-interstitial (I) and their recombination, starting or ending with a host atom S sitting in a regular lattice position.

$$S \rightleftharpoons I + V \qquad (1.14)$$

The heterogeneous ones, also known as Schottky-type of processes

$$S_{surf} \rightleftharpoons I \qquad (1.15)$$

$$S \rightleftharpoons V + S_{surf} \qquad (1.16)$$

are conceived as a kind of single defect injection from an external or internal surface (*surf*) or from an extended defect, that work as point defects sources or sinks.

In a real crystal, the final establishment of equilibrium defect concentration depends on the reaction rates of the processes associated to the defect formation.

Considering a generic defect reaction

$$A + B \underset{k_r}{\overset{k_f}{\rightleftharpoons}} C \qquad (1.17)$$

which could well correspond to the formation/dissociation of a Frenkel pair,

---

[20] Light irradiation or irradiation with, electrons or energetic ions is a convenient way to create non-equilibrium defects.

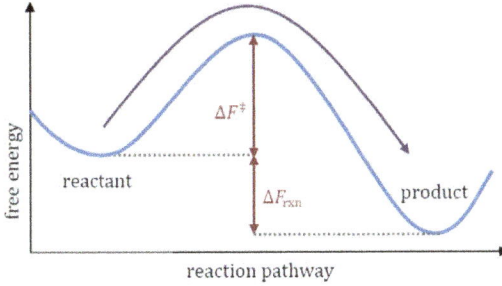

Fig. 16 Configurational energy diagram for the chemical reaction of eq. 1.17. ΔF_{rxn} corresponds to the energy difference between the initial and final equilibrium stage of the reaction pathway.

where k_f and k_r are the *rate constants* of the forward and reverse reaction, we can write for the rate of concentration change of the different components

$$\frac{dA}{dt} = -k_f[A][B] + k_r[C] \qquad (1.18)$$

$$\frac{dB}{dt} = -k_f[A][B] + k_r[C] \qquad (1.19)$$

$$\frac{dC}{dt} = k_f[A][B] - k_r[C] \qquad (1.20)$$

Since at equilibrium the forward and reverse reaction rates are equal, putting one of the above time derivatives to zero, we obtain

$$\frac{[C]}{[A][B]} = \frac{k_f}{k_r} = K \qquad (1.21)$$

where the ratio of the *rate constants* [21] of the forward and reverse reactions corresponds to the equilibrium constant K of reaction (1.17), and corresponds, also, to the Helmoltz free energy of reaction ΔF_{rxn} (see also Fig. 16) *via* the equation

$$K = A exp - \frac{\Delta F_{rxn}}{kT} \qquad (1.22)$$

In turn, the kinetics of the process could be modelled, as an example, with the collision theory, that assumes that for a reaction to occur, reactant species must collide with a

---

[21] for which generally holds $k_f \neq k_r$

minimum energy of collision ΔF^+ called the activation energy, which represents the height of the barrier that should be overcome to bring to success the reaction process.

The Frenkel processss

As shown before, the Frenkel process, supposed to be common in diamond, Si, Ge and grey Sn, could be formally written as

$$S \rightleftharpoons I + V \tag{1.14}$$

where S is a semiconductor atom in an undisturbed lattice position and $I+V$ is a couple of self-interstitials and vacancies, called the Frenkel pair.

Here we assume that the Frenkel pair consists of a couple of individual (free) defect species, without excluding (see below) their reciprocal interaction (and bonding) for close inter-distances. For this reason, the enthalpy of formation of the *I-V* pair could include a term of interaction energy.

Accordingly, in the last case, the formation of a couple of individual (free) defect species would imply the thermal dissociation of the pair.

The atomistic mechanism of the homogeneous process is a transfer of a semiconductor atom from a lattice site to a neighbor interstitial position leaving a vacant site. The equilibrium conditions are reached when the vacant site, the interstitial and the matrix rearrange structurally until both defects and the matrix reach their thermodynamically stable configuration of a close or dissociated pair. After dissociation of the pair, defects might diffuse in the host matrix leading to their random distribution.

The formation enthalpy of a dissociated Frenkel pair is just the sum of the formation enthalpies of the individual defects in their equilibrium configuration [88], since the details of the atomistic processes associated to their formation are irrelevant in the determination of the process enthalpy (or entropy and Gibbs free energy).

In fact, the Gibbs free energy, the enthalpy and the entropy are functions of state and their values depend uniquely on the initial and final point of the process and not on a specific process path [9], different from the rate constants of defect generation and recombination processes that depend, instead, on the atomistic details of the process.

It is apparent that the equilibrium concentration of vacancies and self-interstitials does not necessarily be equal, as suggests by equation (1.14), since the Frenkel process could be coupled to separate generation/ recombination/interaction processes of vacancies and/or self-interstitials at homogeneous (impurities) and heterogeneous sources/sinks. These last consisting again of surfaces, grain boundaries, and of second phases.

We will see in Section 2.4 that homo-valent and hetero-valent impurities react with vacancies to form vacancy-impurity complexes in silicon, germanium and diamond, with temperature-dependent modifications of the concentration of the unreacted (free) vacancies.

Eventually, if the forward or reverse rates of the Frenkel process are not sufficiently fast at a specific temperature to allow the establishment or the re-establishment of equilibrium conditions, the concentration of defects is determined by the fastest alternative process occurring at that temperature, and the concentration product $[V][I]^{22}$ could be no longer constant.

This issue was discussed, among others, by Brown *et al* [96], who modelled the kinetics of the reverse process of eq. (1.14) i.e. the recombination of delocalized self-interstitials and vacancies in silicon

$$V + I \xrightarrow{\ k_{rev}^F(T)\ } S \tag{1.23}$$

as a three steps process, consisting

- in the diffusion of delocalized defects at the reaction site,
- the annihilation of a vacancy and a self-interstitial in a recombination event following the collision of the two defects,
- in the return of the delocalized silicon atom (i.e. the product of annihilation) to its regular lattice position.

The rate constant of the recombination process $k_{rev}^F(T)$ at constant temperature T could be written, if the process is supposed to be diffusion controlled, with an activation barrier ΔH_{rec} for recombination

$$k_{rev}^F(T) = \frac{4\pi a_r}{\Omega c_s}(D_V + D_I)exp - \frac{\Delta G_{rec}}{kT} = A(T)exp - \frac{\Delta H_{rec}}{kT} \tag{1.24}$$

where $a_r = 1$ nm is the capture cross section for recombination, Ω is the atomic volume (of Si in this case), c_s is the lattice concentration of Si atoms ($= 5 \times 10^{22}$ cm^{-3}), D_V and D_I are the diffusion coefficients of vacancies and self-interstitials, $\Delta G_{rec} = \Delta H_{rec} - T\Delta S_{rec}$ is the free energy of recombination/localization and ΔH_{rec} is the activation enthalpy of the process, *i.e.* the activation barrier height.

The recombination free energy includes enthalpic and entropic contribution for the re-localization of silicon atoms. The enthalpy term has been estimated to hold 1.5 eV while the entropic term ΔS_{rec} is computed to hold -3.46k.

---

[22] Using here for concentration the symbol []

Under this hypothesis, the height of the recombination barrier for the reverse reaction is 1.5 eV, sufficiently high to prevent recombination at least at moderate temperatures. The recombination is, on the contrary, almost complete in the case of Frenkel pair components created by irradiation, since the recombination, here, is ionization-enhanced, according to Emsev *et al* [97].

Since the rate of the forward process depends also on the enthalpies of formation of the vacancy-self-interstitial pair, the rate constant of the forward process is given by the following equation

$$k_{forw}^F(T) = B(T)exp - \frac{\Delta H_f^V + \Delta H_f^I + \Delta H_{rec}}{kT} \qquad (1.25)$$

making the forward process even slower.

The equilibrium constant of the Frenkel process K_F could be easily obtained by recalling that at equilibrium the rates of the forward and reverse reaction are equal $k_{forw}^F(T)x_V x_I = k_{rev}^F(T)x_{VI}$

and thus, that the ratio of the rate coefficients satisfy again the condition $K_{eq} = \frac{k_{forw}}{k_{rev}}$.

Therefore, from equations 1.24 and 1.25

$$K_F = exp - \frac{\Delta G_f^{FP}}{RT} = K^o exp - \frac{\Delta H_f^V + \Delta H_f^I}{RT} \qquad (1.26)$$

where ΔG_f^{FP} is the Gibbs free energy of formation of the dissociated Frenkel pair, K^o includes the entropy of formation ΔS_f^{FP} of the dissociated Frenkel pair, and $\Delta H_f^V + \Delta H_f^I \simeq \Delta H_f^{FP}$ is the formation energy of the dissociated Frenkel pair.

As long as the Frenkel process does occur under equilibrium conditions, we can foresee for the product of the activities of vacancies and self-interstitials the following relationship to hold [98]

$$a_I a_V \simeq x_I x_V = K_F \qquad (1.27)$$

where a_I is the activity of self-interstitials, a_V the activity of vacancies and x_I and x_V are their concentrations, which could be assumed to be equal to their activities, due to the extreme dilution of the defect solution.

Equation (1.27) predicts that the product $x_I x_V$ is constant at constant temperature and that in absence of external sources/sinks for vacancies and/or self-interstitials

$$x_I = x_V = x_I^{eq} = x_V^{eq} = K'^o exp - \frac{\Delta H_f^{FP}}{2kT} \qquad (1.28)$$

Instead, in the presence of external sources/sinks for self-interstitials or/and vacancies, we have for the equilibrium concentration of self-interstitials

$$x_I = \frac{K'^o}{x_V^*} exp - \frac{\Delta H_f^{FP}}{RT} \tag{1.29}$$

and of vacancies

$$x_V = \frac{K'^o}{x_I^*} exp - \frac{\Delta H_f^{FP}}{RT} \tag{1.30}$$

where x_V^* and x_I^* are maintained constant, at constant temperature, by the external source or sink, respectively.

It could be, eventually considered that the rate of bulk generation/recombination process could be faster if it occurs at some local surface or bulk heterogeneities, which behave as catalysts, as it happens in conventional chemical processes. In this case the process is only diffusion, not reaction controlled.

An additional issue should be mentioned before concluding the discussion of the Frenkel process, since in literature it is, in fact, supposed to occur with the formation of a pair consisting of a *bound* couple of a vacancy *V* and a self-interstitial *I*.

Centoni *et al.* [68] suggest that it is not unreasonable to suppose, at least in silicon, that a couple of vacancy and a self-interstitial lying at a very short inter-distance should behave differently from isolated defects.

They show, in fact (see Table 4) that the calculated formation energies of the pair depends on the configuration of the self-interstitial in the pair (either tetrahedral or split 100) and are slightly lower (0.2-0.5 eV) than the sum of individual energies (depending on the interstitial configuration and charge), indicating the presence of a non-negligible defect interaction when a Frenkel pair is generated.

Table 4. Interdistances and enthalpies of formation of bound pairs and individual defects in silicon [68].

geometry	r (A)	ΔH_f
VI_T	8.83	7.18
VI_{100}	8.21	7.31
$V^o + I_T^{++}$	∞	7.39
$V^o + I_{100}^{o}$	∞	7.44

Eventually, the stability of a Frenkel pair consisting of a couple of bound vacancies and self-interstitials is even more evident in irradiated diamond, where, thanks to the ability of carbon to form sp^2, sp^3 and π-bonds, stable reconstructed Frenkel pairs, with vacancies and self-interstitials lying few atomic distances apart, are easily generated. The calculated formation energy of this Frenkel pair is worth approximately 12eV [99], quite lower than the sum of the formation energies of separate defects (see Table 5), indicating that in diamond a Frenkel pair consisting of a couple of bound vacancies and interstitials is the stable configuration.

The Schottky processes

Two equilibrium processes could be conceived as responsible for the independent, heterogeneous generation/annihilation of vacancies and self-interstitials, known in literature as Schottky processes.

The atomistic mechanism of the generation of a vacancy involves the removal of a semiconductor atom from the bulk of the phase and its transfer to the surface, where it keeps the configuration of a reconstructed surface atom and leaves a vacant bulk site that, assumes its equilibrium configuration.

As already seen, this process could be formally written as

$$S \rightleftharpoons V + S_{surf} \tag{1.16}$$

where S is a semiconductor atom in a normal lattice position and S_{surf} is a semiconductor atom located at the surface[23] but actually at the surface of any kind of heterogeneous sink or source.

---

[23] This process in the forward direction is associated to an increase of the number of atoms at the surface

The second, responsible for the generation/recombination of self-interstitials, occurs again at the surface /interface of a bulk phase, or at an extended defect (as an example, a dislocation) and consists in the removal of a semiconductor atom from the surface of a bulk phase and its transfer to an interstitial position of the material, where it assumes its equilibrium configuration and could be written as

$$S_{surf} \rightleftharpoons I \tag{1.15}$$

Both processes could occur as equilibrium processes, in competition with the Frenkel process, ruled by the Gibbs free energies of formation of the individual vacancies ΔG_f^V and self-interstitials ΔG_f^I.

As before, the activation energy for the process rate includes a re-localization entropy and enthalpy terms in addition to the Gibbs free energy of formation of the defect, and the equilibrium constant of reaction (1.16) could be written as

$$\frac{a_S^{surf} a_V}{a_S^b} = \gamma_S^{surf} a_V \approx \gamma_S^{surf} x_V = K^o exp - \frac{\Delta H_f^V}{RT} \tag{1.31}$$

where $a_S^b = 1$ is the (thermodynamic) activity of semiconductor atoms in the bulk, $a_S^{surf} = \gamma_S^{surf} x_s^{surf} = \gamma_S^{surf}$ is the activity of semiconductor atoms S at the surface, γ_S^{surf} is the activity coefficient of semiconductor atoms at the surface[24], with $\gamma_s^{surf} = const \neq 1$, and $a_V = x_V$ due to the extreme dilution of the defect solution. By routine, the pre-exponential contains the formation entropy of a vacancy and ΔH_f^V is the enthalpy of formation of isolated vacancies.

The equilibrium concentration of vacancies x_V is eventually given by the following equation

$$x_V = K^{o'} exp - \frac{\Delta H_f^V}{RT} \tag{1.32}$$

where the pre-exponential includes both a term of entropic character and the surface activity coefficient.

In analogous manner, the concentration of self-interstitials generated following eq. (1.15) is given by the following equation

$$x_I = K^{o''} exp - \frac{\Delta H_f^I}{RT} \tag{1.33}$$

where ΔH_f^I is the enthalpy of formation of isolated self- interstitials.

---

[24] Since the surface is a 2D Gibbs phase, we have introduced an activity coefficient term also for the activity of surface atoms, that is intrinsically different from that of the bulk.

The overall equilibrium, however, would foresee the simultaneous satisfaction of the condition (1.29) or (1.30), if the Frenkel process would occur without kinetic hindrances.

Relatively little is experimentally known about the generation/recombination kinetics of point defects occurring in equilibrium conditions, as most knowledge on these processes has been obtained from irradiation experiments of silicon and germanium, where it is supposed that defects (vacancies and self-interstitials) originate from a point source from which they diffuse radially [100].

These experiments showed that recombination of vacancies and self-interstitials and the vacancies interaction with impurities play a major role on the survival of defects and that only a few percent of generated vacancies form stable defects [100]. It has also been shown that metastable Frenkel pairs are not observed in silicon irradiated with electrons or γ-rays, contrary of the case of Ge, where these pairs could be detected and are therefore stable [101]. From platinum profiles in metal diffusion measurements carried close to full equilibrium conditions it was shown that the generation and recombination of Frenkel pairs between 700 and 850 °C is ineffective [98] while Zn-diffusion measurements demonstrate that an entropic barrier exists for the vacancies-interstitials recombination even at 942 °C [102], indicating also the possibility of different structures of vacancies at low and high temperatures

Eventually, slow bulk recombination rates are at the origin of the inhomogeneous distribution of defects in silicon ingots naturally cooled from the growth temperature, since the individual defect recombination could preferentially occur at the ingot surfaces, where reactions are faster than in the bulk of the sample [103], leading to an inhomogeneous, radial and axial, distribution of defects in a silicon ingot [86].

Keeping in mind all previous arguments, we suppose that only at temperatures well above at least to 950 °C the Frenkel process in silicon could be considered analogous to the autoprotolysis of water, capable to keep under control the equilibrium concentration of vacancies and self-interstitials as isolated defect species.

2.3 Experimental determination of defect properties and comparison with theoretical calculations: preliminary issues

2.3.1 Defect characterization

The practical application of the concepts presented in the former sections concerning native point defects in group IV semiconductors[25] is limited by severe difficulties associated to the experimental determination of their concentration in equilibrium conditions, since, as we have seen in Section 2.2.1, neither vacancies nor self-interstitials could be easily detected by spectroscopic analysis in non- irradiated samples.

This condition could even bring us to the conclusion that defects are eminently non-equilibrium species, but luckily enough, the concentration of vacancies could be directly measured by reacting (titrating) them, as an example, with hydrogen

$$V + H \rightleftharpoons VH \qquad (1.34)$$

with the formation of a stable, IR-active VH complex (see Section 2.4) which could be then quantitatively detected with optical absorption measurements [104]. With the same experiments a formation energy of the vacancy in silicon was estimated to hold 4 eV, not far from the theoretical values reported in Table 5.

Vacancies and self-interstitials could be, also, quantitatively studied with self-diffusion (SD) [84,105] or impurity-diffusion experiments [102], these last occurring, however, under non-equilibrium conditions.

These methods will be discussed in details in Section 2.5, but the rationale of SD (and foreign metal diffusion) measurements will be anticipated here, to preliminary enlighten their main features.

The SD experiments are based on the assumption that the diffusion of a semiconductor isotope is mediated by vacancies and/or self-interstitials, in their different charge charges states k and y, under the hypothesis that direct exchange processes play a minor role.

Therefore, the self-diffusion coefficient D^{SD}, when both defects are involved, is given by the following relationship

$$D^{SD} = D_V + D_I = \sum_k f_{V^k} x^{eq}_{V^k} D_{V^k} + \sum_y f_{I^y} x^{eq}_{I^y} D_{I^y} + D_{exch} \qquad (1.35)$$

where f_{V^k} and f_{I^y} are correlation coefficient accounting for the ratio of non-random diffusion of the metal and the random diffusion of the defects in their respective charge

---

[25] As well in the totality of semiconductors

states, $x_{V^z}^{eq}$ and $x_{I^y}^{eq}$ are the concentrations[26], expressed in terms of molar fractions of vacancies and self-interstitials, D_{V^z} and D_{I^y} are the diffusion coefficients of the defects in their respective charge states and D_{exch} is a term associated to the contribution of a direct exchange process, occurring without the intermediate of defects, about which we will discuss in Section 2.4.

The isothermal diffusion profiles of a species i, independently of the experimental technique used, are accurately accounted for by the numerical solution of the Fick's second law

$$\frac{\delta C_i}{\delta t} = D_i^{SD} \frac{\delta^2 C_i}{\delta x^2} \tag{1.36}$$

and the D_i^{SD} values measured at different temperatures are eventually fitted by an Arrhenius plot to determine the activation energy of the self-diffusion process.

It is apparent that in the hypothetical case of measurements conducted in a temperature interval where the mobility of one single defect, in one single charge state [27], is predominant, and the contribution of direct exchange could be entirely neglected[28], the result of an isothermal measurement is the product of

$$D^{SD} = f_D x_D^{eq} D_D \tag{1.37}$$

where x_D^{eq} is the equilibrium concentration of the defect D involved in the diffusion process and D_D is its diffusion coefficient.

As remarked by Bracht et al [102], the main limit of self-diffusion measurements is that even in the case of a single defect contribution in a single charge state, one could only determine the product of the equilibrium concentration and diffusivity of a defect, with the need of an additional experimental or theoretical support to complete the analysis and identifie the nature of the defect and of its formation energy.

Despite these limits, self-diffusion measurements give a deep insight on defect generation and recombination processes, other ways impossible to measure.

As an example, self-diffusion measurements confirm the theoretical forecasts of Car et al [65] that a single defect dominates the diffusion process in the case of germanium (as well as of diamond).

---

[26] which measures the probability of finding a native defect next to a diffusing metallic atom.

[27] This is the case of silicon [84] which will be discussed in the next section.

[28] We will see that this is not the case of silicon.

In fact, as shown in Fig. 17 [106] the experimental self-diffusion values are well fitted with an Arrhenius plot with single activation energy Q

$$D^{SD} = D^{o} exp - \frac{Q}{kT} \qquad (1.38)$$

which confirms the predominance of a single defect in the self-diffusion process.

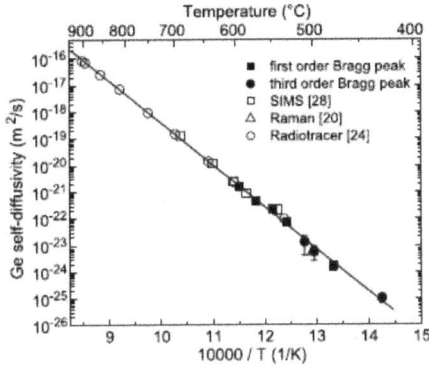

Fig. 17 Experimental temperature dependence of the self-diffusion in germanium. After E. Hüger et al [106]. Reproduced with permission from AIP Publishing LLC, License Number 4020321023101, License date Jan 01, 2017.

Then, from the slope Q of the Arrhenius plot

$$D^{SD} = A exp - \frac{\Delta H_f^D + E_D^*}{kT} \qquad (1.39)$$

one can obtain the sum of the enthalpy of formation of the defect ΔH_f^D and of the activation energy for the defect mobility E_D^*, while the entropic terms of the process are included in the pre-exponential term.

It is apparent that eq.1.37 and 1.39 are applicable for the determination of the equilibrium concentration of the defect x_D^{eq} only if the activation energy for its mobility E_D^* could be independently evaluated or calculated.

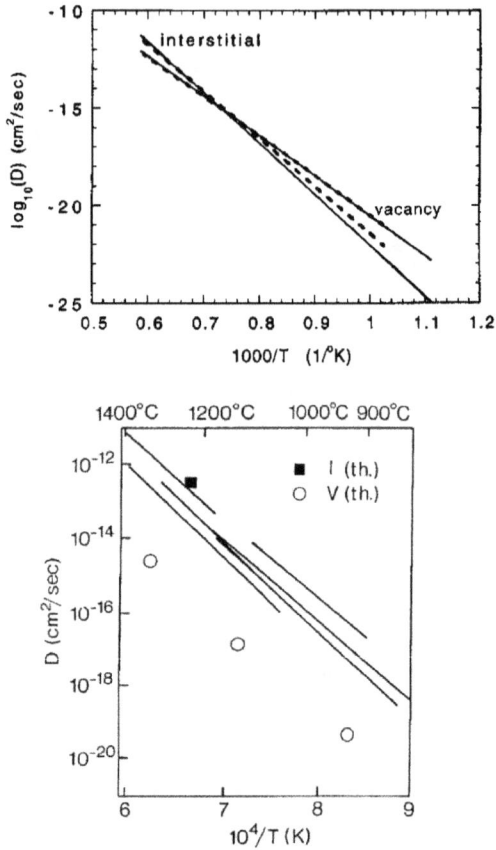

Fig. 18 Calculated values of the temperature dependence of the self-diffusion coefficients of vacancies and self-interstitials in silicon (a) After M. Tang et al [107]. Reproduced with permission from the the American Physical Society, License number 4020851347091, License date Jan 02,2017; (b) after P-E. Blöchl et al [108]. Reproduced with permission from the American Physical Society, License Number 4022610401661, License date Jan 05, 2017.

The case of silicon is more complicated, as will be shown in details in Section 2.5.

In fact, theoretical predictions indicate that both vacancies and self-interstitials mediate the self-diffusion process in silicon, although with different calculated features[29], as it is seen in Fig. 18 [107-108] which displays the calculated diffusion coefficients of self-interstitials and vacancies in silicon, making the interpretation of the experimental results even more challenging.

Since the properties of a defect determined by a self-diffusion experiment are those of a defect submitted to all possible chemical interactions with the other components of the system (a single crystal phase potentially containing several different impurities at specific concentrations)[30], a further problem could arise when one compares self-diffusion results obtained from physical systems not entirely equivalent (i.e. from samples of a semiconductor grown with different techniques, leading to different impurity contents[31]) or with results obtained from theoretical calculations.

In fact, the possible interaction of defects with metallic or non-metallic impurities unintentionally added or arising from any kind of environmental pollution, will lead to different experimental values of the activation energy Q, of the formation enthalpy and equilibrium concentration of defects, biasing results of sample comparison attempts and making improper also their comparison with results of theoretical studies, which rely to ideally clean and perfect structures.

This issue has been recently examined and experimentally tested by Südkamp and Bracht [105], in the attempt to reduce the potential interference of impurities segregated at the surface of the samples on the measured values of self-diffusion coefficient, causing excess defect injections. Details on the technique used and on the success of the initiative will be discussed in Section 2.5, but it could be anticipated here that surface impurity gettering resulted in a substantial improvement of the SD-measurements.

It seems, however, a consolidated habit in the semiconductor research to consider almost uninfluential impurity interaction effects in SD measurements, also in view of the use of extremely pure semiconductor samples, especially those prepared using the heterojunction technique (see Section 2.5).

---

[29] In the case of Tang's calculations [107], self-interstitial contribution is prevailing at high temperatures while interstitials prevail at all temperatures according to Blöchl [108].
[30] And with dislocations, if the phase is not dislocation free.
[31] We have already seen that, as an example, the C and O content of CZ- or FZ-Si samples is very different.

2.3.2 Defect properties as results of theoretical calculations

The tremendous progress of *ab initio* and Molecular Dynamics (MD) calculations about the defect formation thermodynamics and defect mobility which occurred in the last three decades [32] [9, 62, 65, 68, 71, 81, 95, 107,108-114] allows, today, to predict with increasing accuracy the formation energies of defects and the nature of the dominant defect in a specific temperature interval.

We report in Table 5 a selection of recent data concerning the calculated values of formation enthalpies of *neutral* vacancies and self-interstitials[33], from which one could observe, at first, that the calculated values of enthalpies of formation of vacancies and self-interstitials decrease from diamond to germanium, following the increase of the lattice spacing and bond lengths (see Tables 1 and 2).

One can also remark that the calculated values are significantly close to each other, almost independently of the technique used, when referred to the same defect in the same host.

---

[32] The author assumes the responsibility of a personal choice of references, which is far from being exhaustive.
[33] Both vacancies and self-interstitials in doped semiconductors behave as charged species, with charge values and signs depending on the Fermi level.

Table.5 Enthalpy of formation of point defects in diamond, silicon and germanium.

Material	Defect	Enthalpy of formation (eV)	Method	References	References
Diamond	Vacancy	6.52-7.10	First principles	1	71
Diamond	Vacancy	5.96-7.62	DFT	1	71
Diamond	Vacancy	7.2	Local density calculations	2	9
Diamond	Selfinterstitial split (100)	12.3	AIMPRO	3	81
Diamond	Selfinterstitial split (100)	16.7	AIMPRO	2	9
Diamond	BC selfinterstitial	15.8	AIMPRO	2	9
Diamond	Selfinterstitial T_d	23.7	Local density calculations	2	9
Silicon	Vacancy	4.1	Ab initio car-Parrinello	4	108
Silicon	Selfinterstitial	3.3	Ab initio car-Parrinello	4	108
Silicon	Vacancy	3.97	TMD	5	107
Silicon	Vacancy	3.69	DFT	6	68
Silicon	Vacancy	4.3-4.4	*ab initio HSE*	7	114
Silicon	Vacancy	3.38	MD- Car-Parrinello	8	91
Silicon	Interstitial T	4.06	DFT	6	68
Silicon	Interstitial H	3.79	DFT	6	68
Silicon	Interstitial split	3.73	DFT	6	68
Silicon	Interstitials split	3.7	TBMD	5	107
Silicon	Interstitial split	3.49	DFT	9	109
Silicon	Interstitial split	3.84	DFT	10	95
Silicon	Interstitial T	4.07	DFT	10	95
Silicon	Interstitial H	3.80	DFT	10	95
Silicon	Interstitial caged	3.85	DFT	10	95
Silicon	Frenkel pair	7.39	ab initio	6	68
Germanium	Interstitial split	3.55	DFT	9	109
Germanium	Interstitial T	3.85	DFT	9	109
Germanium	Interstitial H	3.99	DFT	9	109
Germanium	Vacancy	2.2-2.4	ab initio CGA	11	111
Germanium	Vacancy	2.9	ab initio HSE	12	113

It should further be noted that the average value of the enthalpies of formation of vacancies (3.4-4.1 eV) in silicon is only slightly higher than that of the most stable self-interstitials, the split I_{100} one (3.3-3.8 eV), allowing to predict that both vacancies and self-interstitials should play a comparable role in self-diffusion and metal-diffusion processes, which is in good agreement with the results reported in Fig. 18. It is also possible to conclude that the split I_{100} and the hexagonal interstitial in silicon are the most stable configurations of self-interstitials in silicon.

The calculated formation energy of the vacancy in germanium is, instead, much lower than in silicon, and also lower than that of self-interstitials. We expect, therefore, that vacancies should be the predominant defects in germanium.

Eventually, the difference between the defect formation energies is extreme in diamond, where the average formation enthalpy of self-interstitials is about 15 eV larger than that of vacancies, leading to the prediction that vacancies should be the prominent defects also in single crystal diamond.

It could be assumed that the qualitative reason of the very high formation energies of self-interstitials in diamond stems from the high density of the diamond lattice and from the stiffness of the diamond bonds (see Tables 1 and 2), which impede bond twisting and the large lattice relaxation needed for the formation of self-interstitials.

As an example, while the relaxation[34] associated to the formation of the T_d interstitial is only around 5%, it grows up to 15% and 20% in the case of the split and bond-centred (BC) interstitials, bringing to higher relaxation energies, and lower formation energies. However, their energies of formation remain so high to preclude to self-interstitials a significant contribution to self-diffusion in diamond [9].

2.4   Interaction of impurities with native point defects

Defect-impurity interactions could occur during all crystal growth and device fabrication processes, as well as at the temperature of self-diffusion or metal- diffusion processes, potentially leading to defect complexes formation, which either shift the equilibrium defect concentration under the simultaneous occurrence of the Frenkel process or bring to novel equilibrium concentrations.

Therefore, the chemistry of defect interactions plays a key role in several semiconductor processes, as is the case of the segregation of excess oxygen and carbon impurities in semiconductor silicon during thermal annealing, which is mediated by defects [8], or that

---

[34] Relaxation leads to lower energies of formation.

of the hydrogenation and gettering processes, where defect-impurity interactions do influence or even determine the final success of these processes.

Defect-impurity interactions, ruled by their Gibbs free energies of reaction (and by strain, if the volume of the product of the interaction would not fit well with the semiconductor host lattice), must be supposed to systematically occur in synthetic semiconductors grown from a melt, like silicon and germanium, even when grown from an extremely pure feedstocks and in extreme clean conditions.

In fact, the ingots are easily contaminated by the gaseous environment of the growth furnace and/or by pick-up of impurities from the crucible walls, which contaminate the melt at ppm or ppb amounts, depending on the impurity, and thereafter, also the solid phase [8].

Although the melt growth of synthetic semiconductors provides substantial impurity rejection for impurities contaminating the feedstock that present segregation coefficients $k = \frac{x_M^s}{x_M^l} \ll 1$, where x_M^s is the concentration of the impurity M in the solid and x_M^l is the concentration of the impurity M in the liquid, the residual impurity concentration in the solid[35] is or might be sufficiently high to affect the optical and electrical properties of the semiconductor and, possibly, the concentration of defects *via* chemical interaction processes.

As an example, the residual oxygen and carbon content is typically $> 10^{18} cm^{-3}$ and $> 10^{17}$ cm^{-3} respectively, in Czochralsky (CZ) silicon ingots and $\leq 10^{16} cm^{-3}$ for both impurities in Float Zone (FZ) silicon. The average metal impurity content in CZ silicon is, instead, higher than $10^{18} cm^{-3}$. High purity germanium (HPGe) used for detector applications is much purer, with an average impurity concentration of $<10^{10} cm^{-3}$ [115], the main impurities being B, Ga and P.

Depending on the growth conditions, also Ge could be heavily contaminated with oxygen ($\approx 5.5 \ 10^{17} cm^{-3}$) [116].

Eventually, natural diamond is systematically contaminated with nitrogen and traces of silicon, and defect-impurity reactions could occur during its geothermal formation, while synthetic diamonds grown with Chemical Vapour Deposition (CVD) methods could be contaminated with traces of metals[36] and hydrogen.

---

[35] As an impurity profile is present in every melt grown ingot, the impurity concentration on top could be so high to deteriorate its electronic properties.
[36] Metal contamination by CVD or plasma-CVD processes is a well know problem in the industrial production of hepitaxial silicon.

The chemistry of impurity-defect interactions can be discussed and experimentally investigated by assuming the possible occurrence of three different processes.

The first is the direct formation of an impurity-defect $D-M$ complex

$$D+M \rightleftharpoons D-M \qquad (1.40)$$

with defects D *(vacancies and self-interstitials)* and impurities M randomly distributed in the semiconductor matrix.

The thermodynamic success of this process depends on the Gibbs free energy of reaction ΔG_{react}

$$K = \frac{[D-M]}{[D][M]} = exp - \frac{\Delta G_{react}}{kT} \qquad (1.41)$$

For $\Delta G_{react} \ll 0$ the reaction will be almost fully split on the right and the complex concentration will be close to the impurity concentration if [D]>>[M] or to the defect concentration if [M]>>[D].

Oxygen and carbon are the impurities of main interest for reactions involving vacancies, considering their ubiquitous presence in silicon and germanium, while nitrogen plays the same role in natural diamond.

The process described by eq. 1.40 should occur jointly to the Frenkel and Schottky processes in the temperature ranges where these processes are fast, consequently determining the concentration of the unreacted defects.

The second and the third potential processes, instead, are either a true equilibrium between vacancies and metal impurities

$$M_s \rightleftharpoons M_i + V \qquad (1.42)$$

or between self-interstitials and metal impurities

$$M_s + I \rightleftharpoons M_i + S \qquad (1.43)$$

which could sit in both substitutional M_s and interstitial positions M_i in the diamond lattice, where S is an atom of the host crystal.

The process under eq. 1.42 takes the name of Frank-Turnbull (FT) or dissociative [117], while that under eq. 1.43 takes the name of Gösele or kick-off mechanism [118], respectively.

The mechanism of this last process is conceived, in fact, as the kick-out of an impurity sitting in a substitutional position under the impact of a self-interstitial. In literature, with the unique exception of the paper under [119] equilibrium (1.43) is systematically, but erroneously, written with the reversible reaction $M_s + I \rightleftharpoons M_i$, which does not satisfy the mass conservation law.

We will see that these processes are proposed as the elementary steps of foreign metal diffusion processes (see Section 2.6) involving few metals, called hybrid metals, whose diffusion is mediated by defects.

Defect complexes could be detected by Fourier Transform Infra-Red (FTIR), ESR, DLTS, Laplace DLTS, Raman, PAS and ENDOR measurements, making the experimental studies on them especially well affordable in irradiated semiconductors, where the defect concentration is orders of magnitude higher than that of the equilibrium.

2.4.1 Vacancy-impurity complexes formation in silicon, germanium and silicon-germanium alloys

Vacancies are particularly prone to complex defects formation, due to the presence of DBs, which could share their unpaired electrons in reconstructed covalent bonds with impurities of the IV, V and VI group, with relevant influence on semiconductor properties when their presence is associated to deep levels.

Among these impurities, interstitial oxygen O_i is significantly reactive with vacancies in Si and Ge, leading to the formation of a thermodynamically stable VO complex species (the calculated binding energies E_B range between 1.4 and 1.8 eV, see Table 6), known as the A centre [120-124] that is electrically- (acceptor centre at E_c-0.17 eV) and spectroscopically-active (see Table 6).

The structural configuration of the A centre is shown in Fig. 19a. One sees that the VO complex is a system of four dangling bonds partially saturated by two oxygen bonds, supposed to be generated according the following reaction

$$V + O_i \rightleftharpoons VO \qquad (1.44)$$

or, alternatively

$$O_i \rightleftharpoons VO + I \qquad (1.45)$$

in the hypothesis that the vacancy-self-interstitial equilibrium would be ruled by a Frenkel process.

In both cases the concentration of "free" vacancies should be considerably reduced.

The properties of the A center were mostly studied in samples irradiated with high energy ions, electrons or γ rays, to enhance its concentration, but it is actually present also in non-irradiated Si and Ge samples. It, in fact, is known to enhance the oxygen diffusivity at temperatures below 300 °C, where the interstitial oxygen is practically immobile [123], involving a mechanism of formation/dissociation/oxygen transfer *via* the VO species [125] that favors the oxygen diffusion.

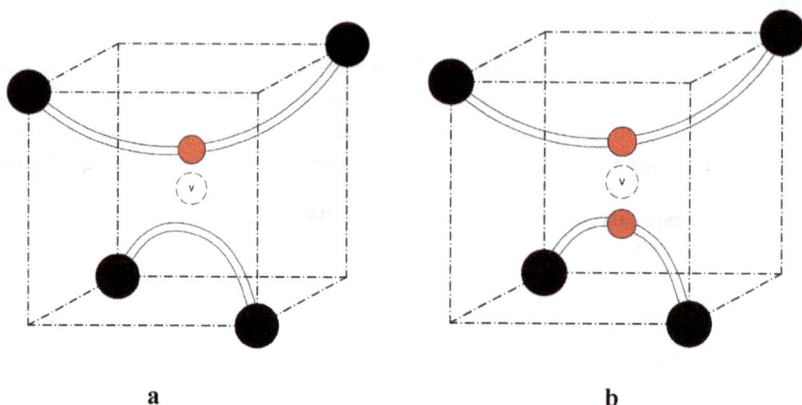

a   b

Fig. 19 Schematic representation of the VO (a) and VO₂ complexes (b) Adapted from a suggestion of C.A. Londos et al [125].

The physical, structural and thermodynamic properties of the A centre have been the subject of several experimental and theoretical studies [120-123, 126-132] since very recently [133-134], which showed, also, some key features of its behaviour.

The A centre, in fact, converts first to VO_2, whose structure and physical properties are reported in Fig.19b and Table 6, and then to more oxygenated VO_n species[37], with n=3,4,5, with a maximum temperature limit for their stability lying around 750 °C according to Claeys [135], but the chemistries of these transformations have been under question until recently.

---

[37] The VO_n species look as a chemical nonsense. One shoud label V_mO_n, these more oxidized (n>2) vacancy species to satisfie stoichiometry rules. For an easier comparison with literature data we will however use the VO_n term in the following of this section, considering *n* a fractional number.

According to Litvinov *et al* [130], in fact, VO anneals out at 120-140 °C in germanium, above 300 °C in Si according to Coutinho [121] and Londos *et al* [125] with the formation of VO_2.

According to Monakov and Svensson [74], the increase of the concentration of VO_2 during the thermal annealing of VO is proportional to the decrease of concentration of VO, and supports the idea that VO and VO_2 are two equilibrium species, and that the reaction of formation of VO_2 can be given by the following equation

$$VO + O_i \rightleftharpoons VO_2 \qquad\qquad (1.46)$$

The Monakov's conclusion is supported by the work of Coutinho *et al* [121], who suggest that the interaction of two interstitial oxygens O_i with the formation of a self-interstitial and VO_2 is unlikely

$$2O_i \rightleftharpoons VO_2 + I \qquad\qquad (1.47)$$

as it leads, instead, to the formation of a fast moving dioxygen centre O_{2i}

$$2O_i \rightleftharpoons O_{2i} \qquad\qquad (1.48)$$

which is stable against the formation of a O_2 molecule and of VO_2.

However, the one step saturation of all the four DBs of a vacancy

$$V + 2O_i \rightleftharpoons VO_2 \qquad\qquad (1.49)$$

cannot be entirely neglected, at least in principle, since its calculated formation energy (2.76 eV) is higher by about 1 eV than that of the reaction 1.47 (see Table 6).

Therefore, it would be reasonable to conclude that reactions 1.46 to 1.49 do occur in parallel under equilibrium conditions, as we will see at the end of this section as the outcome of the Rye-Larsen [134] work.

It should be noted, however, that not only does the literature fail to provide a common view on the chemical processes involved in the VO_2 formation, but also that the calculated values of the enthalpy of formation of VO_2 should be taken with some caution.

In fact, the calculated value of the enthalpy of formation of VO_2 from VO +O (1.7 eV) (see Table 6) correspond almost exactly (1.7-1.83 eV) to the value of the enthalpy of formation of VO, corresponding to the saturation of the first two DBs of an empty vacancy, while in order to accommodate two oxygen atoms in the same volume, we expect the onset of a repulsive interaction between the two oxygen atoms [125] and a

51

shortening of the bond lengths of the oxygen atoms in VO_2, in good agreement with the increase of the vibration frequency that is in fact, experimentally observed (see Table 6).

Some further light is provided by the experimental results of Chroneos et al [136-137] carried out using FTIR spectroscopy and reported in Fig. 20, that shows that VO begins to transform in VO_2 above 300 °C, while VO_2 starts to decompose at T> 500 °C, demonstrating that VO and VO_2 are simultaneously present in a rather large temperature range.

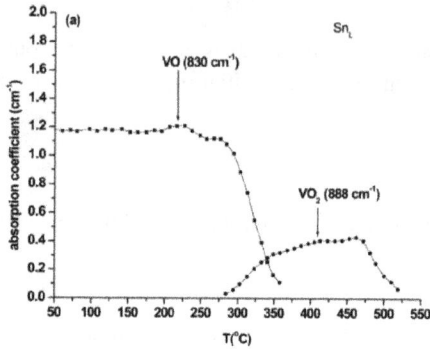

Fig.20 The thermal evolution of VO and VO_2 defects. After A. Chroneos et al [137]. Reproduced with permission from AIP Publishing LLC, License Number 4020871153291, License date Jan 02, 2017.

Eventually, the temperature dependence of the equilibrium concentration of the VO and VO_2 complexes together allow that free vacancies can be evaluated using eq. 1.45 - 1.46 and 1.49 and, the respective equilibrium constants K_{VO}, K_{VO_2} and $K_{VO_2}^*$

$$K_{VO} = \frac{[VO]}{[O_i][V]} \qquad (1.50)$$

$$K_{VO_2} = \frac{[VO_2]}{[VO][O_i]} \qquad (1.51)$$

$$K_{VO_2}^* = \frac{[VO_2]}{[V][O_i]^2} \qquad (1.52)$$

taking for the initial concentration of vacancies that arising from a Schottky equilibrium

$$[V] = K_s^\circ exp - \frac{\Delta H_S}{kT} \qquad (1.53)$$

In turn, the equilibrium constants can be calculated using literature data for binding energies or enthalpies of formation of VO (ΔH_f^{VO}) and of VO_2 $\Delta H_f^{VO_2}$ or $\Delta H_{VO_2}^*$)

$$K_{VO} = K_{VO}^\circ exp - \frac{\Delta H_{VO}}{kT} \qquad (1.54)$$

$$K_{VO_2} = K_{VO_2}^\circ exp - \frac{\Delta H_{VO_2}}{kT} \qquad (1.55)$$

$$K_{VO_2}^* = K_{VO_2}^{*\circ} exp - \frac{\Delta H_{VO_2}^*}{kT} \qquad (1.56)$$

Table 6 Experimental and calculated values for the enthalpies of formation of VO and VO_2 complexes (LVM = Local Vibration Modes). References under [].

Element	Enthalpy of formation of VO (eV)	Enthalpy formation of vacancies (eV)	Enthalpy of formation of VO_2 (eV)	Enthalpy of formation of GeVO and SnVO (eV)	LVM of VO (cm-1)	LVM of VO_2 (cm-1)
Silicon	1.7 ±0.4 [122]	3.3 [122]	2.7[122] (V+2O)		835 [120]	895 [120]
Silicon	1.6 [121]	3.61[121]	1.7 [74]		830 (VO°) 877(VO⁻) [74]	889[74]
Silicon	1.85[123] theor 1.86[123] exp		(VO+O) 1.73[121]			
Germanium	1.86 [121]	2.20 [121]	(VO+O) 1.30 [122]		621.4-669.1 [130]	731,771,801 [130]
Sn-Doped Si	1.30 [137]			1.26 [137]		
Ge-Doped Si	1.30 [139]			1.56 [139]		

A typical result of this kind of calculations is displayed in Fig. 21, for the concentration of VO and VO_2 centres in a virtual silicon sample containing an oxygen concentration of 10^{18} cm^{-3} [122], using for the enthalpy of formation of vacancies, of VO and of VO_2 (via the process V+2 0) the values reported in table 6.

Under the hypothesis that VO_2 arises directly from the saturation of all its DBs in a single step (eq.1.49), the results show that vacancies trapped as VO and VO_2 complexes are thermodynamically stable from 700 to 1200 °C, with VO concentrations C_{VO} well comparable with the experimental values of vacancies concentrations obtained with foreign metal diffusion experiments, showing their almost complete oxidation to vacancy-oxygen complexes.

The same figure shows also that the calculated concentration of free vacancies C_V (dotted line) is systematically below that of trapped vacancies in the entire range of temperatures investigated.

The apparent disagreement with the experimental Chroneos results displayed in Fig.20, that show that VO_2 is, in fact, the dominant species already above 400 °C, comes from the hypothesis that VO_2 arises from the direct oxidation of the vacancy.

The chemistry of VO and VO_2 complexes is apparently influenced also by the presence of isovalent impurities (C, Sn and Ge), that react with vacancies giving rise to complexes that compete in stability with the VO complex [137-139] and do influence as well its transformation to VO_n species.

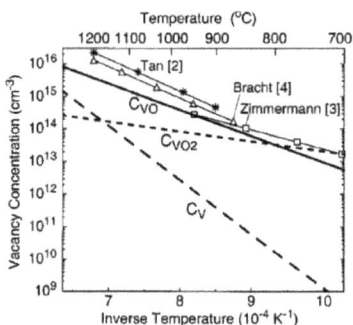

*Fig.21 Calculated equilibrium concentration of free vacancies and VO and VO_2 complexes as a function of temperature. After A. Casali et al [122]. Symbols display the experimental values of equilibrium concentration of vacancies obtained from impurity diffusion experiments by Δ [102]; □ [98] * [228]. Reproduced with permission from the AIP Publishing LLC. License number 4020880569931. License date Jan 02, 2017.*

The case of Sn is particularly interesting [136], as it forms a SnVO complex in silicon with a calculated formation energy (1.26 eV) comparable to that of the VO complex (see Table 6), whose formation is, in fact, strongly depressed in the presence of Sn, as it is seen in Fig.22.

It is eventually known that the diffusion of all shallow donors (P, As, Sb, and Bi) and shallow acceptors in silicon and germanium is as well vacancy assisted [140] and that the formation of stable vacancy complexes with shallow donors (the E centres), formed by capture of a mobile vacancy in correspondence of site occupied by a substitutional group V species, is obtained by electron or γ-rays irradiation [141].

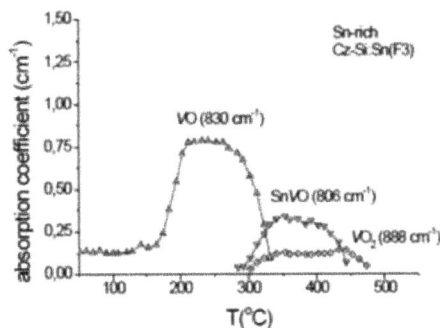

Fig.22 The thermal evolution of the VO, VO$_2$ and SnVO defects in a Sn-doped Si sample. After A. Chroneos et al [136]. Open access Journal.

To conclude the analysis of vacancy-oxygen complexes in silicon and germanium it is worth to remark that the partial or total saturation of the DBs of the vacancy implies the partial or total removal of the intrinsic attributes of the vacancy. Furthermore, it results in the formation of localized Si-O bonds in the lattice, which could potentially work as precursors or nucleation sites for the oxygen aggregation/precipitation processes[38], including the formation of thermal donors (TDs), which occurs at least in parallel with the evolution of the vacancy-oxygen complexes in the same temperature range (350-500 °C).

---

[38] For a comprehensive treatment of oxygen precipitation processes in silicon see [229] A.Borghesi, B.Pivac, A.Sassella, A,Stella (1995) Oxygen precipitation in silicon *J.Appl.Phys.* **77** 4169-4244

This issue has been discussed, among others, by Rava *et al* [142] who concluded that the formation of vacancy-oxygen complexes constitutes the initial stage of thermal donors formation, while Ourmazd *et al* [143] and then Chadi [144] and Gregorkievicz and Bekman [145] concluded that the two processes seem to proceed without reciprocal interferences, without excluding, however, a vacancy core for a thermal donor center [145].

Eventually, Rye-Larsen [134] studied the evolution of VO_n complexes generated by MeV electron irradiation of carbon-loan CZ Si and of thermal donors during sequential annealing at 450°, 500° and 550 °C, using FTIR absorption measurements. (see Fig. 23, 24 and 25)

It can be seen that during the annealing at 450 °C the VO_n family evolves with the initial disappearance of VO and the formation of VO_2 (see Fig.23 a) and that the formation of high-n VO_n complexes occurs after a long anneal. It can also be seen that TDs coexist with vacancy-oxygen complexes only after 400 h of annealing (see Fig.23b), showing that the TDs formation requires a long induction period.

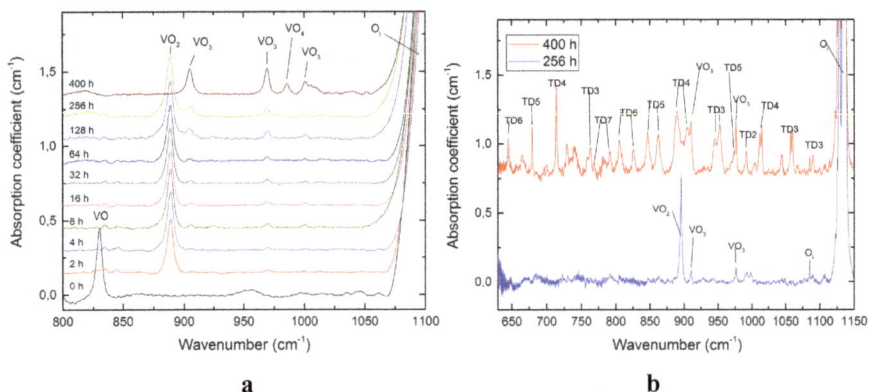

Fig. 23 Time evolution of VO_n complexes (left) together that of thermal donors (right) in MeV electron- irradiated samples during annealing at 450 °C [134]. Reproduced with the kind permission from Mette Fjelltveit Rye-Larsen, Materials, Energy and Nanotechnology Dept., University of Oslo.

The situation changes for TDs at 500 °C (see Fig. 24a), since the evolution of V-O complexes occurs with a faster rate, but the formation of TDs (see Fig. 24b), is inhibited in the irradiated samples (lower curve, sample A500i) while it occurs in the case of the non-irradiated samples after 81 hr.

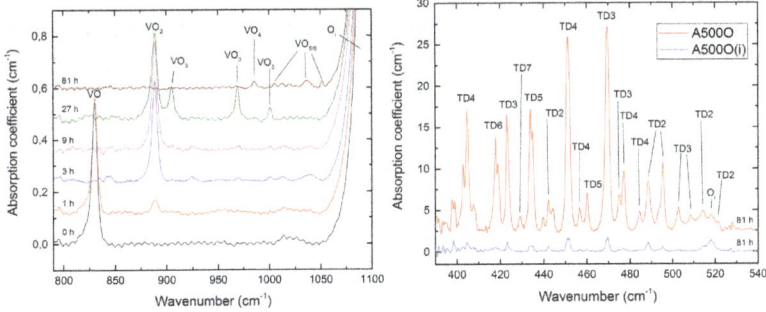

Fig. 24 Time evolution of VO_n complexes (left) and of thermal donors (right) in MeV electron- irradiated samples during annealing at 500°C [134]. Reproduced with the kind permission from Mette Fjelltveit Rye-Larsen, Materials, Energy and Nanotechnology Dept., University of Oslo.

Eventually (see Fig.25), thermal dissociation effects at 550°C seem to be dominant for both VO complexes and TDs, with the IR signatures of the latter which could be no more detected after prolonged annealing.

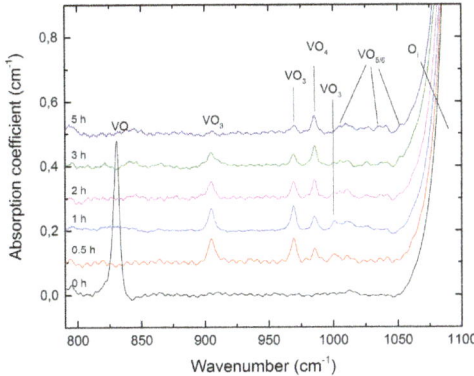

Fig.25 Time evolution of VO_n complexes in MeV electron- irradiated samples during annealing at 550 °C [134] Reproduced with the kind permission from Mette Fjelltveit Rye-Larsen, Materials, Energy and Nanotechnology Dept., University of Oslo.

The experimental evidence that VO_n complexes and TDs coexist at 450 °C after prolonged (400 h) annealing, but that the formation of TDs is inhibited by the presence of VO complexes at 500 °C seem to indicate that these oxygen complexes do not evolve without reciprocal interferences, as concluded by former researchers.

Apparently, V-O complexes behave as sinks for some precursor species of TDs, as could be the case of the dimeric O_{2i} species, although the work of Coutinho [121] indicates that oxygen dimers are stable against V-O. This would imply some reconsideration of the issues concerning the formation of the VO complexes discussed at the beginning of this Section.

2.4.2. Vacancy-carbon complexes in silicon

C-V, P-V and As-V pairs are also stable in both Si and Ge, with calculated binding energies which increase from C to As (see Table 7) [146]. As suggested by the author of ref. 146, the absolute values could be questioned, not the trend that indicates an increasing stability of these vacancy pairs in silicon.

We could therefore expect that carbon contamination could affect the self-diffusion coefficient in FZ silicon, considering that its concentration ($\leq 10^{16}$ cm^{-3}) has the same order of magnitude of vacancies concentration.

Table 7. DFT-calculated binding energies of vacancy pairs in Ge and Si [146].

Defect pair	E_b(Ge) (eV)	E_b(Si) (eV)
C-V	0.07	0.36
P-V	0.52	1.23
As-V	0.60	1.34

2.4.3 Vacancy-impurity complexes in diamond: The Nitrogen-Vacancy centre

Atomic size and lattice spacing constraints make the structure of the vacancy-impurity complexes in diamond different to that discussed for Si and Ge. In fact, both isovalent- and heterovalent- impurities bonded to vacancies sit in substitutional positions, occupying one of the four corners of a vacancy tetrahedron (see Fig. 26) for the case of the nitrogen- vacancy (NV) complex, thus simultaneously satisfying the bonding needs of the impurity and leaving less than four un-saturated DBs, depending on the chemical nature of the impurity.

As an example, it could be foreseen that the substitution of one carbon atom with nitrogen would leave three unsaturated DBs, while its substitution with O would leave only two unsaturated DBs

Another difference, that makes vacancy-impurity complexes in natural [39]diamond very challenging, is their almost common property to behave as color centres, which attracted the earliest interests on their structural properties and on their optical and magnetic signatures, in view of possible applications in advanced quantum computing machines [147-156].

Among various vacancy-impurity centres in diamond, the nitrogen-vacancy (NV) complex is granted by a more than satisfactory knowledge of its thermodynamic, structural and physical properties. Its study started from the hypothesis that the nitrogen content of natural diamond of the Type 1b, was probably responsible of the formation of a defect centre exhibiting a strong luminescence emission at 637 nm.

However, given the low nitrogen[40] content of natural diamond, the concentration of NV centres generated by spontaneous interstitial-substitutional exchange and vacancy trapping at substitutional nitrogen atoms in the due to geological processes or subsequent thermal annealing is too low to be useful for technological applications, and substitutional nitrogen N_C is the main defect[41].

Their concentration could be, however, enhanced by irradiation with high energy particles, followed by annealing at 600 °C, a temperature at which vacancies generated by irradiation become mobile and are trapped as substitutional nitrogen [157], leading to a random distribution of NV centres in a surface layer of the crystal of a thickness comparable with the irradiation depth.

Using nitrogen implantation, nitrogen could be, instead, precisely positioned with nanometers separation on a diamond surface [158-159] and vacancies produced as by-products of the implantation process are trapped at nitrogen implants as NV centers.

Although most of implanted-nitrogen rests on interstitial positions, a respectable fraction could be incorporated in substitutional positions by replacement collisions [159], leading to a final conversion efficiency from nitrogen to NV ranging around 10% [160].

---

[39] and synthetic diamond
[40] The nitrogen content in Type 1b diamond is around 500 ppm, and 2000-3000 ppm in Type 1a.
[41] With a calculated activation energy for diffusion around 6 eV, which makes it very stable up to 1700-2100 °C.

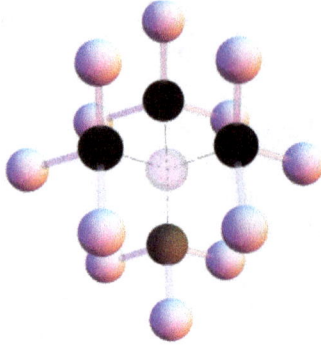

Fig. 26 The structure of the NV complex in diamond; black spheres are C atoms while the red one is a nitrogen atom. After M.W. Doherty et al [161]. Open access Journal, Creative Commons Attribution 3.0 Unported (CC-BY).

This relatively poor yield of the VN formation process in diamond is due to a competition with the formation of divacancy centres in irradiated diamond.

In fact, since the calculated energy gain (4.2 eV) associated to the formation of the divacancy [157]

$$V^o + V^o \rightarrow V_2^o + 4.2eV \qquad (1.57)$$

is higher than that associated to the formation of the NV center

$$V^o + N_s^+ \rightarrow VN^o + h + 3.3eV \qquad (1.58)$$

(see Table 6) the divacancy formation is thermodynamically favored.

Table 6. Formation energies of defects involved in the set-up of the VN complex and enthalpies of formation ΔH_{form} of the NV centre and of the divacancy in diamond [157].

	E_{form} (eV)	ΔH_{form} (eV)
V	7.14	
N	3.96	
NV°	6.21	-4.89
V_2	10.08	-4.2

The calculated structure of the V-N complex in diamond displayed in Fig.26 turns out to a very complex electronic structure, which has been solved only recently [161]. The results of ESR measurements have shown that the electronic states of this complex are highly localized at the vacancy and could be described with a molecular model constructed with a linear combination of molecular orbitals. Its properties are consistent with a six electrons model, of which five belong to the five dangling bonds of the nearest neighbor C and N atoms and the sixth is localized at the vacancy [161].

2.4.4 Vacancy-impurity complexes in diamond: other impurity-vacancy centres

The increasing experience available of synthetic diamond growth stimulated further experimental and theoretical interest to the NV complex as well to different types of impurity-vacancy complexes, which can easily be prepared by MOCVD (Metal Organic Chemical Vapor Deposition) or ion implantations processes and could present interesting optoelectronic applications [162-165], including the fabrication of nanoscale sensors for physical and biological applications [160].

Among impurity-vacancy complexes, also oxygen-vacancy complexes are supposed to be present in diamond films CVD-grown in a 0.1% oxygen atmosphere, in view of their strong PL luminescence at 2.561 eV [166] which well fits with the calculated Zero Phonon Line (ZPL) emission (2.56 eV) [167] associated to the Vacancy-Oxygen centre shown in Fig. 27.

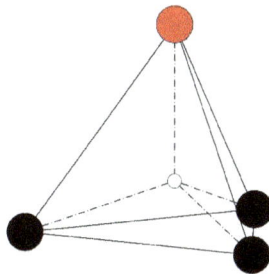

Fig. 27 Structure of the O-V centre in diamond (black spheres carbon atoms, red sphere oxygen atom) [167].

Different from interstitials oxygen atoms in silicon and germanium, oxygen sits in substitutional positions in diamond [168] and in the VO center, like nitrogen in the NV center, and thus saturates two of the four DBs of the vacancy.

As a further example, the GeV center, obtained by Ge implantation at room temperature (RT) or by CVD or MOCVD doping, exhibits a sharp RT luminescence at 602 nm and could work as a single photon source [165]. The structure of this complex, close to that of the SiV centre which emit at 736 nm, was elaborated from first principles and modelled starting from the hypothesis that the GeV centre should have a structure like the NV centre, with a Ge atom substituting one of the four C atoms having the vacancy in the centre, leaving unsaturated all of its DBs.

Since it was found that the Ge atoms in the diamond lattice do prefer to sit in an interstitial position, due to their size larger than that of C, the final structure is that of a divacancy having a Ge atom in an interstitial site between, as it is shown in Fig. 28.

Therefore, the formation of a Ge-V centre occurs according the following reaction

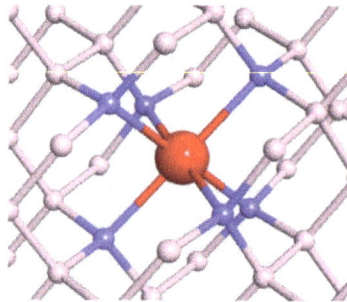

$$V + GeV \rightarrow Ge_iV_2$$
(1.59)

Fig.28 The structure of the Ge-V centre with the Ge atom in red in the centre of a di-vacancy. After T. Iwasaki et al [165]. Open access article.

Also, transition metals-vacancy complexes (TM-V) are known to give rise to ultra-bright emissions in diamond, due non-intentional impurity contamination of synthetic diamond in its growth process. This is the case of CrV complexes in diamond nanocrystals [169] grown on sapphire crystal substrates by microwave enhanced CVD. They exhibit a strong emission centered at 756 nm. As the diamond was not intentionally doped, the emission was attributed to Cr out-diffusion from the substrate, which contained ppm amounts of Cr as an impurity, though the authors do not exclude that the impurity involved could also be Al or O, although the calculated emission from V-O complexes lies around 450 nm and Cr-related centers in diamond exhibit luminescence spectra in the 740-770 nm range [169]. The Ni-V complexes, also commonly found in synthetic diamond are, instead, only electrically active [162].

Interesting properties are, eventually, presented by the N-V-H complexes, whose presence in microwave plasma CVD-grown N-doped diamond single crystal samples has been deduced by Glover et al [170] on the base of the ESR spectrum exhibited by these samples. The same ESR results are consistent with a structure of this complex, where H is directly bound to the nitrogen atom, maintaining the original trigonal symmetry of the NV center.

This attribution has been questioned by Goss et al [171] who mentioned, at first, that the substitution of one C atom neighbor to a vacancy with a N atom removes one dangling bond, satisfying in the meantime the bonding needs of the nitrogen and reducing the total energy of the vacancy, leading to a complex VN with three DBs, instead of four.

On this principle, an interstitial H atom would saturate a second dangling bond, with a further decrease of the total energy of the vacancy, resulting in a complex VNH with two dangling bonds instead of three.

However, theory do not support a model where H is bonded to a N-atom in the NV complex [171] and predicts the presence of a Bond Centred (BC) hydrogen interstitial close to a nitrogen atom that forms a complex with the neighbor C atom, with a little interaction with the nitrogen atom, suggesting that hydrogen would be mostly involved in the saturation of a vacancy DB.

This conclusion seems to be supported by the DFT calculated value [171] of the energy liberated for the hydrogen passivation[42] reaction (6.8 eV)

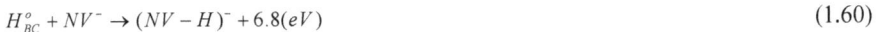

$$H^o_{BC} + NV^- \rightarrow (NV - H)^- + 6.8 (eV) \tag{1.60}$$

where H^o_{BC} is a neutral, bond centered interstitial hydrogen, which closely corresponds to the energy involved in the passivation of a single dangling bond (6.5-6.9 eV) at the (111) diamond surface [172].

---

[42] Here "passivation" is used in analogy of other passivation reactions caused by hydrogen [8].

2.4.5 Vacancy-impurity complexes: a new notation

We have shown in the last four sections that a variety of structural and chemical changes are associated to vacancy-impurity interactions in group IV semiconductors, which strongly differ in the case of Si and Ge, on one side, and diamond on the other side, depending on the different localization of the impurity in the structure of the vacancy-impurity complex.

The difference could be made symbolically apparent if we introduce a variant of the Kröger and Vink notation used so far, assuming that a neutral vacancy in a semiconductor could be formally considered a radicalic S_4^{4*} species, where S is a semiconductor atom at a corner of an empty tetrahedral cavity and star indicates an unpaired electron (a DB).

Therefore, in the case of Si (and Ge) the product of a reaction between an heterovalent impurity M^x (where x is the excess valence electrons over 4) with a vacancy is a $Si_4^{(4-x)*}M^x$ species, while is a $C_3^{(4-x)*}M_C^x$ species in the case of diamond, with M_C^x sitting on one of the corners of the empty tetrahedron, if the atomic size of the impurity is compatible with the atomic size of carbon in the diamond lattice.

If size compatibility is missing, as in the case of Ge impurities in diamond, the product of the interaction is a di-vacancy-Ge center, with the impurity atom in the core of the divacancy (see Fig. 28).

We will see that this notation might be useful in the case of compound semiconductors (SiC and GeC) of the fourth group.

Chapter 3

3.   Self-diffusion experiments and their theoretical modelling as practical tools to deduce nature and presence of native defects in group IV semiconductors

3.1   Experimental approach and outcomes

Self-diffusion in solids can, theoretically, proceed by two different mechanisms, one dominated by defects and another one assisted by a direct exchange (DE) process.

In the first case, the presence of a defect, a vacancy as an example, delivers an easy path, with a limited height of the saddle point (and a limited activation energy) for a jump of an atom from a lattice position to a neighbor, empty, lattice site.

In the second case, the process involves the site interchange of two adjacent atoms *via* subsequent bond breaking and rotation steps in different lattice planes, with a more complex configurational diagram than that of the process mediated by a vacancy.

Albeit there is the theoretical and experimental evidence of a negligible DE contribution to self-diffusion in metals, the situation could be different in covalent solids with directional bonding. In fact, DE is the dominant diffusion mechanism in the basal plane of graphite [173] and could not be ignored in the case of silicon and germanium.

In 1986, Pandey [173] developed a model for the calculation of the activation energy for DE in silicon for a multiple path process involving bond rotation, bond distortion and a minimum of two broken bonds (see its configurational diagram in Fig. 29). He called this process Concerted Direct Exchange (CDE) and showed that it leads to a calculated value of 4.3 eV for the activation energy [173-174], close to the value of 4.7 eV calculated by Car *et al* [65].

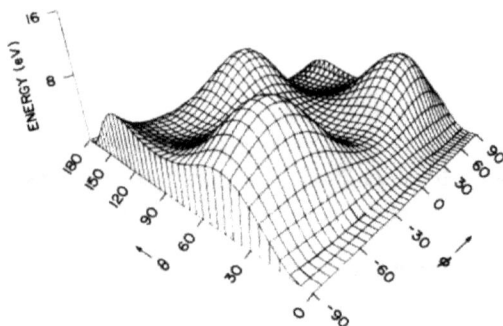

Fig.29 An example of a total energy surface for the ideal concerted exchange After K.C. Pandey [173]. Reproduced with permission from the American Physical Society, License Number 4021351010444, License Date Jan 03, 2017.

This value is, also, comparable with literature values of the activation energies of processes mediated by defects in the case of silicon (4.1 eV at 1100 K and 5.1 at 1500 K), but, according with Ural *et al* [175], there is also the experimental evidence that the CDE mechanism accounts for less than 14% in silicon.

DE counts even less in the case of diamond (see Table 7) [9], while takes a different scheme in the case of germanium diffusion in silicon [176-177], where the calculated activation energy for CDE is only slightly higher (E_a=4.25 eV) than that mediated by self-interstitials (E_a =3.25+0.44=3.69 eV) and vacancies (E_a= 3.63+0.37=4.00 eV), although the DE process itself could be mediated by an energetically low cost intermediate defect.

Table 7 Calculated values of the activation energies (in eV) for processes mediated by defects in diamond and of the direct exchange energy [9].

Vacancy	Tetrahedral interstitial	(100) split interstitial	Bond-centered interstitial	Direct exchange
7.2	23.6	16.7	15.8	13.2

For all these reasons, the Direct Exchange mechanism is accounted as a parasitic path or totally neglected in the analysis of Self-Diffusion and Impurity-diffusion measurements that are the practical tools to get information of the nature, concentration and mobility of defects in a crystalline solid.

As it is well known, self-diffusion or impurity-diffusion measurements are conventionally carried out by depositing, first, on the surface of the sample a layer of an isotope of elemental semiconductor or of the metal which is then diffused into the semiconductor at a convenient temperature. After annealing, the isotope- or the metal impurity-profiles are determined using SIMS (Scanning Ion Mass spectrometry) or other analytical techniques, including DLTS and scanning probe resistivity[43], depending on the tracer. Eventually, the experimental profiles are used to numerically compute the diffusion coefficient from the solution of the Fick's second law at the temperature of measurements.

Several experiments carried out at different temperatures are finally used to display the temperature dependence of the diffusion coefficient and then of the $x_D^{eq} D_D$ term (the transport capacity), under the approximations described in Section 2.3.

A more elaborated technique employs periodic isotope superlattice homo- or heterostructures to obviate the long duration of the self-diffusion experiments, especially at low temperatures [84].

In this case, the diffusion profile is obtained by Raman Scattering or Neutron Reflection[44] techniques.

In agreement with the theoretical arguments discussed in Section 2.3, self-diffusion or impurity-diffusion measurements are expected to exhibit an Arrhenius relationship with a single value of the activation energy Q_X for a process mediated by a single defect X, that could be either a vacancy or a self-interstitial in a single charge state

$$D_i^{SD} = f_X x_X^{eq} D_X = A_X exp - \frac{Q_X}{kT} = f_X exp \frac{\Delta S_X^m + \Delta S_X^f}{k} exp - \frac{\Delta H_X^m + \Delta H_X^f}{kT} \qquad (1.61)$$

where ΔS_X^m and ΔS_X^f are the migration and formation entropies of the defect X, and ΔH_X^m and ΔH_X^f are their migration and formation enthalpies, respectively, writing for the temperature dependence of the defect concentration

---

[43] If the tracer induces electrical cconductivity changes in the semiconductor host.
[44] See the original papers on the isotope heterostructure techniques, as this topic is outside the interest of this Review.

$$x_X^{eq} = exp - \frac{\Delta G_X^f}{kT} = Aexp - \frac{\Delta H_X^f}{kT}$$

(1.62)

where ΔG_X^f is the Gibbs free energy of formation of the defect. According to Blöchl *et al* [108], the formation entropy of the defect S_f^X should present a configurational character related to the volume of the space phase available for the defect.

The condition of a process mediated by a single defect, according eq.1.61, is experimentally proven for diamond, using the ^{13}C isotope [9, 178-179 (see Fig. 30) and for germanium, using isotopically enriched ^{70}Ge/^{nat}Ge [106], (see Fig. 17), considering that theoretical estimates already provided the forecast of vacancy- assisted (see Table 3 and 8) diffusion processes in diamond and Ge.

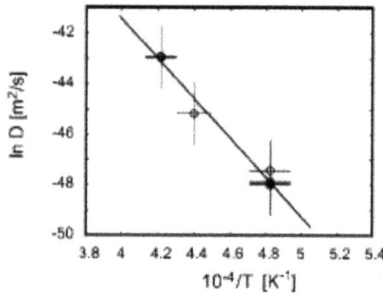

Fig. 30 Arrhenius plot of the carbon self-diffusion coefficient in diamond at 10 GPA. After K.T. Koga et al [178]. Reproduced with permission from the American Physical Society, License Number 4021360531776, License Date Jan 03, 2017.

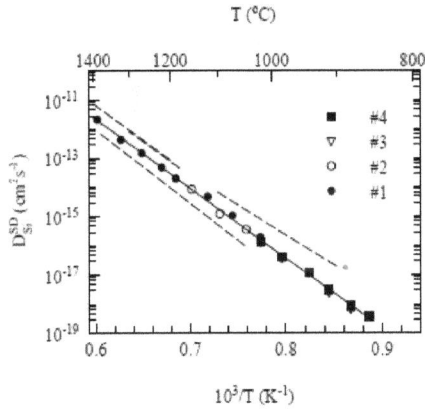

Fig.31 Temperature dependence of the self-diffusion coefficient of silicon, After H. Bracht [84]. Reproduced with permission from the American Physical Society, License Number 4021370070265, License Date Jan 03, 2017.

Different is the case of silicon, that has been hardly debated since recently, since the simultaneous contribution of vacancies and self-interstitials could not be excluded, [84,105,180-181], as already discussed in a former Section.

In fact, see Fig. 31 [84] and Fig. 32 [181] which report the most representative results concerning silicon since 2013, one could observe that if the measurements are carried out within the 855-1388 °C experimental temperature range (Fig.31) [84] a single contribution to the activation energy (E= 4.75 eV) is measured for values of D_i^{SD} spanning over seven orders of magnitude

$$D_{Si}^{SD} = 530 \ exp - \frac{4.75 \pm 0.04(eV)}{kT}(cm^2 sec^{-1}) \tag{1.63}$$

Instead, if the measurements are carried out down to 650 °C (see Fig. 32) a distinct curvature appears, that is not consistent with a process dominated by a single value of the activation energy.

In agreement with theoretical estimates [107] that foresee the contribution of both interstitials and vacancies to the self-diffusion coefficient, and a crossover of self-interstitials over vacancies at about 1080 °C, the experimental results of fig. 32 could be well fitted with the following equation

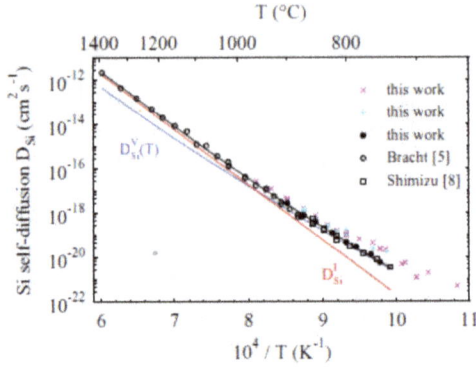

Fig.32. Self-diffusion coefficient of silicon measured with isotopically enriched ^{28}Si in the temperature range 650- 960 $^{\circ}$C. After R. Kube et al [181]. Reproduced with permission from the American Physical Society, License Number 4021371354908. License date Jan 03, 2017.

$$D_{Si}^{SD} = (2175.4exp - \frac{4.95eV}{kT} + 0.001exp - \frac{3.52eV}{kT})cm^2 sec^{-1} \tag{1.64}$$

which accounts for a process dominated by the contribution of both defects, and is also in excellent agreement with the results of Shimizu et al [180], given, in turn, by the following equation

$$D_{Si}^{SD} = (2175.4exp - \frac{4.95eV}{kT} + 0.0023exp - \frac{3.6eV}{kT})cm^2 sec^{-1} \tag{1.65}$$

where the first term accounts for self-interstitials and the second one for vacancies.

However, the association of the second term of equation (1.64 and 1.65) to vacancies is not conclusive, since the pre-exponentials (0.001-0.002 cm^{2}sec^{-1}) are too small to fit with a significant value of the migration and formation entropy of vacancies [181].

It is therefore suggested that the experimental temperature dependence of the activation energy of self-diffusion, which varies practically in a continuous manner from 3.7 eV at 700 °C to 4.7 eV at 1400 °C, could be better explained by assuming temperature-independent thermodynamic properties of self-interstitials and temperature-dependent thermodynamic properties of vacancies, which would eventually mean [181] that vacancies are localized at low temperatures and become extended at high temperatures.

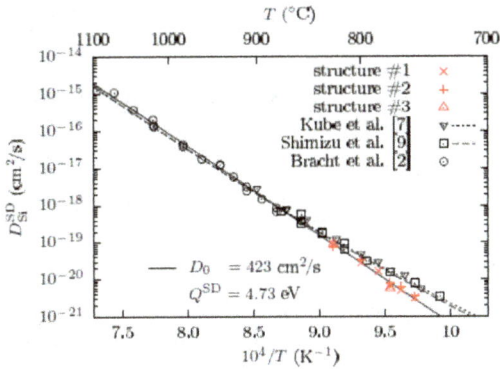

Fig.33 Self-diffusion coefficient of silicon measured with isotopically enriched ^{28}Si in the temperature range 755 and 825 °C. After T. Südkamp et al [105]. Reproduced with permission from the American Physical Society, License Number 4021380618259, License Date Jan 03, 2017.

This topic has been recently re-considered by Südkamp and Bracht [105] on the hypothesis that the bowing observed at low temperatures could be due to unwanted impurities adsorbed at the surface of the sample (OH- groups or transition metals including copper) whose dissolution (with silicides formation, in the case of transition metals) would be associated to the generation (injection) of excess defects, and thus, to the enhancement of the silicon diffusivity, especially at low temperatures[45].

Impurity gettering could be, therefore, adopted to remove or, at least, to minimize these parasitic processes by suppressing the impurity pick-up from the sample container walls.

To this scope, self-diffusion measurements were carried out within 755 and 825 °C enclosing the samples in quartz ampoules filled with crushed silicon, used as a getter.

The comparison of the results of these measurements reported in Fig.33 with the former ones displayed in Fig.32 shows that with the use of a crushed silicon getter the experimental self-diffusion data can be extremely well fitted with an Arrhenius plot extending within 9 orders of magnitude, with a single diffusion enthalpy, see eq.1.66.

---

[45] At low temperatures the equilibrium concentration of defects is so low that could be easily incremented by external sources.

$$D_{Si}^{SD} = 423 \ exp \ -\frac{(4.73 \ \pm \ 0.02)eV}{kT} cm^2 sec^{-1} \tag{1.66}$$

Consequently, also the previous hypothesis of a temperature-induced change of the size of the defects seems to be inappropriate.

However, also the hypothesis that a metallic impurity (copper, as an example, that is a systematic contaminant of single crystal silicon[46]) could be responsible for the deviation from the linearity of the Arrhenius plot observed at low temperatures, and that its gettering by crushed silicon restores its linearity, deserves some attention.

In fact, the proof of copper gettering from crushed silicon was only indirectly obtained using resistivity measurements in germanium samples[47] carried-out with- and without silicon gettering.

In addition, while copper gettering in germanium, using silicon as the getter, is thermodynamically effective, and therefore beneficial, due to the larger solubility of copper in silicon [182]. In the temperature range of the Südkamp and Bracht's measurements, copper gettering in silicon using crushed silicon would only work in the case of silicon samples heavily contaminated with the metal, a very odd condition.

Furthermore, Cu-gettering would imply an intimate connection of the hypothetically Cu-rich sample surface with the getter, to allow an efficient solid/solid mass transfer that is not obvious in the case of a flat sample surface in occasional contact with a powdered getter.

Eventually, and independently of the nature of the defect which dominates the self-diffusion process at low temperatures, it could be excluded that silicon gettering would be beneficial for the suppression of bulk processes involving defect-impurity interactions, leading to thermodynamically-stable species at low temperatures.

Instead, the suppression of metal vapors or oxygen pick-up from the container walls is certainly activated using a mass of crushed silicon.

---

[46] This topic has been recently discussed by the present author in [8].
[47] Cu concentration changes in germanium could be easily monitored by electrical conductivity measurements.

3.2 Theoretical modelling and outcomes

While it is demonstrated that self-diffusion measurements represent an excellent mean to get information of the $x_X^{eq}D_X$ product, with important feedbacks for technological processes, at least an independent measurement[48] or a theoretical estimate is necessary to establish the nature of the defect involved in the diffusion process and, then, the separate contribution of migration and defect concentration terms.

The results of numerical evaluation of defect formation energies are reported in Table 4, while Table 8 displays the results of Molecular Dynamics (MD) and *ab initio* calculations concerning the activation enthalpy of self-diffusion mediated by a specific defect (see for general considerations about this topic [108,183].

One might observe the excellent fit of the experimental and calculated activation energies for self-diffusion in the case of diamond and germanium, where a single defect is involved.

The case of silicon is more complicated, as vacancies and self-interstitials are both involved, and the two processes occur simultaneously without evident crossover [105]. Still, also for silicon, the agreement between calculated and experimental data is satisfactory and confirms the prevalent contribution of vacancies in self-diffusion processes carried out at low temperatures.

---

[48] As an example, the impurity diffusion experiments is discussed here in Section 3.2.

Table 8 Experimental and calculated values of activation energy of self-diffusion in diamond, silicon and germanium.

Element	Tracer	Temperature range (°C)	Activation energy of self-diffusion (eV)	Defects involved	Activation energy of single defect diffusivity (eV)	Ref.
Diamond	^{13}C	2075-2375	6.8 ±1.6 (exp)	Vacancy		178
Diamond			6.71±0.20 (calc)	Vacancy		179
Silicon	^{29}Si ^{30}Si epi on FZ Si	855-1388 °C	4.75±0.04 (exp)	Interstitials Vacancies	4.95 (exp) 4.14 (exp)	84, 184
Silicon	$^{29}Si/^{28}Si$	650- 960		Interstitials Vacancies	4.95 (exp) 3.52 (exp)	181
Silicon	^{28}Si epi on CZ Si	755-825 with getter	4.73±0.02 (exp)	Interstitials vacancies	4.82±0.05 (exp) 4.65±0.05 (exp)	105
Silicon	$^{28}Si/^{30}Si$	735-875		Interstitials vacancies	4.95 (exp) 3.6(exp)	180
Silicon				Interstitials Vacancies	5.18(calc) 4.07 (calc)	107
Silicon			4.03 (calc)	vacancies		185
Germanium	^{71}Ge	549-890	3.14±0.92 (exp)	vacancy		186
Germanium	^{70}Ge	429 -904 °C	3.13±0.03 (exp)	vacancy		106
Germanium	^{71}Ge	822-1163K	3.14±0.02 (exp)	vacancy		186
Germanium			3.17 (calc)	Vacancy		187

3.3 Foreign metal diffusion as an additional method to deduce the nature and properties of defects

It was shown in the previous section that self-diffusion measurements give valuable information on the transport capacities $x_I D_I$ and $x_V D_V$ of self-interstitials and vacancies in semiconductors, using tracer species chemically equivalent to the atoms of the matrix. Diffusion, in this case, does occur in the practical absence of a chemical potential gradient (IUPAC definition) and, therefore, in almost complete equilibrium conditions, such to exclude any interference of the diffusing species on the mobility and the concentration (or, better the thermodynamic activity) of defects.

Foreign metal diffusion could give additional information on defect properties, if either vacancies or self-interstitials, or both, mediate the diffusion process, but with several severe criticalities, which we will also discuss in this section.

Most metals are not suitable to this scope because they are intrinsically fast diffusers, i.e. diffuse as interstitial species without the mediation of point defects.

Some metals, as Au, Pt, Zn, S in silicon (see Fig.34) and Ir, present, instead, the desired properties [85, 98, 102, 119, 188-193] since they exhibit an hybrid substitutional-interstitial (s-i) character, being practically immobile when sitting in substitutional position (M_s) and mobile when sitting in interstitial sites (M_i).

Copper in (dislocated) Ge [117] exhibits the same s-i property, while Ir in silicon couples a (minor) contribution of intrinsic interstitial diffusion to a defect-mediated diffusivity [192-193].

They represent a transition from the purely (fast) interstitial diffusivity of H, Cu, Ni and Fe to the (slow) diffusivity of dopants (B, P, Sb, As), Si and C.

All these si metals present, also, the property that their equilibrium solubility in substitutional positions x_M^s is systematically larger (or much larger) than the interstitial one x_M^s. As an example, in the case of Zn in Si, the ratio is $\frac{x_M^i}{x_M^s} = 0.01 - 0.03$ [102-119].

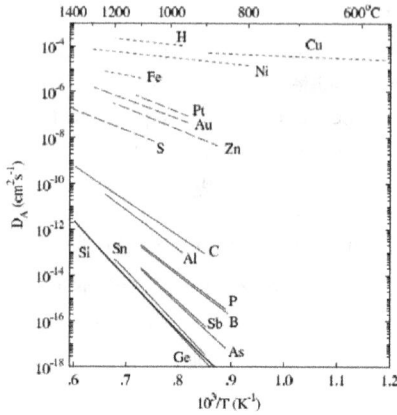

Fig. 34 Temperature dependence of diffusion coefficients of metallic impurities in silicon. Fast diffusers are labeled with short dashed lines, hybrid diffusers with long dashed lines and solid lines are used for slow diffusers. After H. Bracht [85]. Reproduced with permission from CRC Press, Order Details ID70226034; License ID 4023570103278, Confirmation Number: 11616679, Order Date: 01/07/2017.

The key issue that makes these metals suitable to give direct information on the formation energies of defects and, thus, on their equilibrium concentrations, is the defect capability to mediate their transfer from substitutional to interstitial lattice sites.

Since all these metals are spontaneously transferred from a metallic phase, properly[49] deposited at the surface of the semiconductor sample, to the semiconductor lattice as mobile interstitial species Me_i [102-119], it could be assumed that they accommodate in the preferred equilibrium substitutional positions either *via* the production of self-interstitials

$$Me_i + S \rightarrow Me_s + I \tag{1.67}$$

---

[49] The mass amount of metal deposited on the surface of the sample should be sufficient to remain present at constant unit activity during the entire duration of the process. It may consist of a liquid metal source, as is the case of Zn, which is liquid at the diffusion temperatures in Si. In this case Zn vapors refresh continuously the injection capacity of the sample source.

(where S is a lattice atom), thus inducing a local self-interstitials supersaturation, or *via* the absorption of vacancies

$$Me_i + V \rightarrow Me_s \qquad (1.68)$$

and the generation of a local vacancy undersaturation, when the solubility of the foreign metal is larger than the equilibrium concentration of self-interstitials or/and vacancies $x_M^{eq} > x_V^{eq}, x_I^{eq}$ [50], and the transport capacity of the metal is larger than that of defects $D_M \gg D_I, D_V$. Otherwise, the equilibrium concentrations of self-interstitials and vacancies will be rapidly recovered by defect out-diffusion or indiffusion from the surface [102].

Equilibrium conditions at the metal (α)/semiconductor phase (β) $\mu_M^\alpha = \mu_M^\beta$ imply, in fact, that the semiconductor phase is locally saturated with the metal at its equilibrium solubility x_M^{eq}, and that the total $x_M = x_M^s + x_M^i$ concentration of the metal coincides with its (equilibrium) solubility value.

In the absence of intrinsic interstitialcy diffusivity, the defect-mediated diffusion involves the sole migration of metal interstitial species Me_i with a process which requires, to be effective, a substitutional/interstitial site exchange of the metallic species, mediated either by self-interstitials (kick-off mechanism) [118, 188]

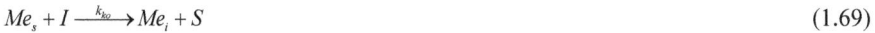

$$Me_s + I \xrightarrow{k_{ko}} Me_i + S \qquad (1.69)$$

where S is again a lattice atom and k_{ko} is the forward rate constant of the kick-off process, or by vacancies (Frank & Turnbull mechanism)

$$Me_s \xrightarrow{k_{FT}} Me_i + V \qquad (1.70)$$

where k_{FT} is the forward rate constant of the FT process.

In turn, since in thermodynamic equilibrium conditions the forward and reverse rates of the kick-of and of the FT reactions are balanced[51], K_{ko} is the equilibrium constant of the kick-off process

$$\frac{x_M^i}{x_M^s x_I^{eq}} = \frac{k_{forw}}{k_{rev}} = K_{ko} \qquad (1.71)$$

---

[50] That is the general case, with the exception of Ir.
[51] In the physical literature this is called the principle of detailed balance.

and K_{FT} is the equilibrium constant of the FT process

$$\frac{x_M^i x_V^{eq}}{x_M^s} = \frac{k_{forw}}{k_{rev}} = K_{FT} \tag{1.72}$$

Different from self-diffusion processes, impurity diffusion occurs under a chemical potential gradient that is the thermodynamical driving force of the process.

In the case of defect-assisted hybrid metal diffusion, however, the impurity migration process is correlated to the local concentration excess/defect of self-interstitials or of vacancies, whose concentration profiles within the semiconductor samples depend only on the diffusion coefficients of self-interstitials or of vacancies.

The local concentration excess/defect of self-interstitials or of vacancies controls, in fact, the reaction rates of the kick-off or the FT processes.

It could be also argued that after sufficiently long diffusion times, local equilibrium conditions for the defect concentration are reached ($x_V = x_V^{eq}$; $x_I = x_I^{eq}$), which could also be expressed by the following equation for a kick-off process [102]

$$D_I^* x_{M_i} = D_I x_I^{eq} \tag{1.73}$$

and

$$D_I^* x_{M_i} = D_V x_V^{eq} \tag{1.74}$$

for a FT process, where D_I^* is an effective diffusion coefficient [102]

The fit of metal diffusion profiles for a self-interstitial- mediated process can be done with the numerical solution of the set of following differential equations [119]

$$\frac{\delta \tilde{x}_s}{\delta t} = k_b \, x_I^{eq} \left(\tilde{x}_i - \tilde{x}_s \tilde{x}_i \right) \tag{1.75}$$

$$\frac{x_i^{eq}}{x_s^{eq}} \frac{\delta \tilde{x}_i}{\delta t} = D_I^* \frac{\delta^2 \tilde{x}_i}{\delta \chi^2} - \frac{\delta \tilde{x}_s}{\delta t} \tag{1.76}$$

$$\frac{x_I^{eq}}{x_s^{eq}} \frac{\delta \tilde{x}_I}{\delta t} = D_I^* \frac{\delta^2 \tilde{x}_I}{\delta \chi^2} + \frac{\delta \tilde{x}_s}{\delta t} \tag{1.77}$$

(and a similar set for a vacancy-mediated process) where the subscripts s and i are equivalent to M_s and M_i, respectively and all the $\tilde{x}$ terms are normalized concentrations to the equilibrium concentration $\tilde{x}_z = \frac{x_z}{x_z^{eq}}$, with z= M_s, M_i and I, applying the minimum number of fitting parameters which include the equilibrium constants (1.71) and (1.72) and their temperature dependence.

A necessary precondition to extract from the results of the numerical solution the properties of self-interstitials and vacancies is an excellent previous knowledge of the solubility of the metal and of its temperature dependence.

The results of the impurity diffusion measurements, carried out in a wide range of temperatures, lead to exponential relationships of the transport capacities and then of the defect concentrations. This is the case of the Zn-diffusion measurements carried out by Bracht *et al* [102] displayed in Fig.35, where also a comparison with the literature data is presented.

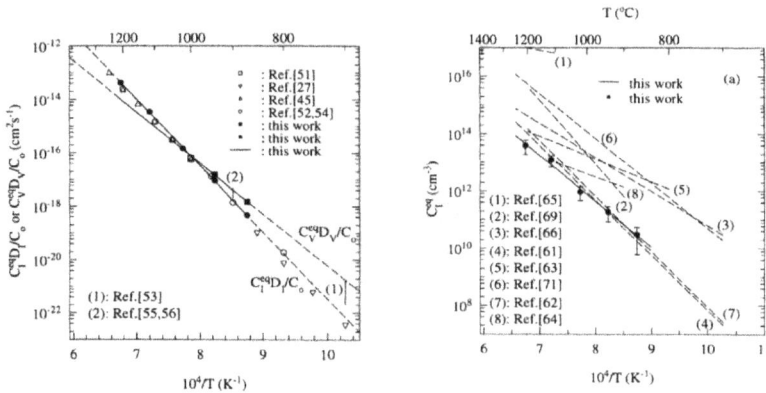

Fig. 35 Temperature dependence of the transport capacities (left) and of the self-interstitials concentration(right) (black symbols and full lines obtained from Zn-diffusion in comparison with literature data, After H. Bracht et al [102]. Reproduced with permission from the American Physical Society, License number 4018880982357, License date Dec 30, 2016.

Here, as we have seen for silicon self-diffusion, both self-interstitials and vacancies mediate the diffusion process, leading to a coupling of the kick-off and FT reactions and to a transport capacity of the impurity which brings the contribution of both defects [119]. The corresponding numerical values of the formation enthapies of self-interstitials and vacancies are reported in Table 9, which reports also the results relative to different measurements carried out using Au [98], Ir [193] and Pt [98, 189, 191].

One can observe from Table 9 that the experimental values of formation energies of vacancies are systematically in defect of at least 1 eV with respect to the experimental value arising from positron annealing spectroscopy (3.6 eV), which fit well also with the average value arising from theoretical modelling, amounting to $\approx 4 eV$ (see Table 4).

The experimental values of formation energy of self-interstitials are, instead, closer to their theoretical values, which amount to ≈ 3.8 eV, as it seen, again, in table 4.

Table 9. A selection of available values of enthalpy of defect formation ΔH_f and activation energy of diffusion ΔE_D^* determined with foreign-metal diffusion measurements in silicon.

Substrate	Metallic impurity	Mobile defect	Temperature range (°C)	ΔH_f (eV)	ΔE_D^* (eV)	References
Si FZ, undoped	PAS (*)	V	RT-1175	3.6 ±0.2		194
Si FZ, B-doped	Ir	V	875-1050	2.44 ± 0.15	1.66± 0.16	193
Si FZ, undoped	Zn	I	870-1208	3.18± 0.15	1.77± 0.12	102
Si FZ, undoped	Zn	V	870-942	2.0	1.8	102
Si FZ, B doped	Au	I	800-900	3.835 (**)	0.965 (**)	98
Si FZ, B doped	Pt	V	700-950	1.162 (**)	2.838 (**)	98, 189
Si FZ, B doped	Pt	I	700-950	3.835 (**)	0.965 (**)	98,189
SiCZ and Si FZ	Pt	V	680-842	2 (**)	2.03 (**)	191
SiCZ and SiFZ	Pt	I	680-842	4.4 (**)	0.44(**)	191

(*) Result obtained with positron annealing spectroscopy.
(**) simulation parameters used to fit the metal diffusion.

It is hard to reach an unambiguous conclusion about the causes of the significant dispersion of the experimental values of the vacancy properties and of the relative good fit of the experimental properties of interstitials with their theoretical values, though few could be suggested.

As a first issue, the high temperatures at which impurity diffusion processes are carried-out would favor any kind of environmental interaction thermodynamically and kinetically possible within the system under measurement. Since vacancies behave as highly chemically reactive species, their experimental formation energies could depend on their interaction with impurities (different from the metal diffusing species) or other defects, and thus, deviate from the theoretical values.

However, their averaged experimental values over a sufficient number of samples of common origin, not only have a practical value for their application to technological processes but should be considered true (effective) properties of the materials involved in the actual practice, following the Coulson's comment reported in the Introduction and the arguments discussed in the Preface of this Book

Self-interstitials, due to their minor or negligible chemical reactivity, should present behaviour closer to the theoretical one.

This averaging procedure presents, however, different problems to overcome, before being effective.

Silicide formation in the case of Au and Pt diffusion could add, as an example, the additional problem of defect recombination at heterogeneous sinks, and effects of excess defects injection at the Me/ semiconductor interface, which has been tried to be minimize by selecting area coverages as low as possible (10^{15}at cm^{-2}). However, at these low coverages the boundary conditions for Pt and Au atoms are thermodynamically poorly defined [191].

Then, the selection of crucial model parameters, as the metal solubility, is certainly one of the most direct causes of large deviations from effective values, if we consider, as an example, that the equilibrium solubility of Pt in substitutional c_s^{eq} and interstitial positions c_i^{eq} in the silicon lattice, used by Jacob et al [191] are given by the following equations

$$c_s^{eq} = 6.4x10^{29} exp - \frac{3.45eV}{kT} \; ; c_i^{eq} = 4.8x10^{19} exp - \frac{1.7eV}{kT} \tag{1.78}$$

while hold

$$c_s^{eq} = 4.4x10^{24} exp - \frac{2.212eV}{kT} \; ; c_i^{eq} = 1.79x10^{22} exp - \frac{1.920eV}{kT} \tag{1.79}$$

for Zimmermann and Ryssel [189] leading to a difference of the order of 0.8 eV in the formation energy of vacancies and of 0.5 eV in the formation energy of self-interstitials, as could be seen in Table 9.

Eventually, the reason why the best results are obtained with Ir and the worst with Pt is not easy to be understood, unless mentioning that the solubility of Ir is so close to the equilibrium concentration of vacancies [193], and that the metal diffusion process occurs in the presence of a negligible deviation from the vacancy equilibrium concentration.

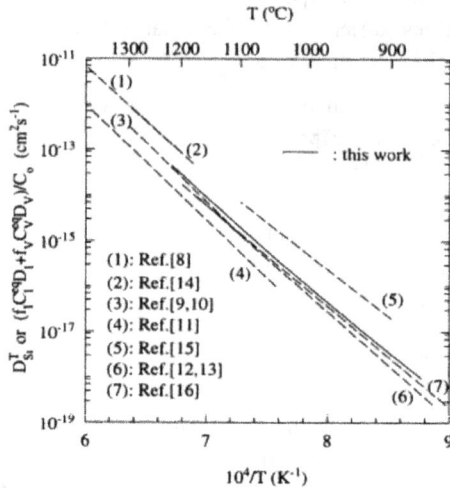

Fig. 36 Temperature dependence of the self-diffusion coefficient (solid curve) in comparison with transport capacities results obtained with Zn-diffusion measurements in silicon. After H.Bracht et al [102]. Reproduced with permission from the American Physical Society, License Number 4018860506445, License date Dec 30, 2016.

The case of Zn deserves particular attention, since different from noble metals, Zn itself is chemically reactive and could interact with oxygen, with the formation of ZnO complexes which are photoluminescence (PL) active [195] and with the VO complex, forming a ZnVO complex. The thermodynamic stability of these complexes is unknown, but we cannot, a priori, exclude that the equilibrium concentration of vacancies could be buffered by dissociation equilibria involving Zn- complexes.

Since these equilibria could not be involved in self-diffusion measurements, and the $D_i x_I$ products obtained with Zn-diffusion experiments are in excellent agreement with those obtained with self-diffusion measurements [102] (see Fig. 36), one can exclude the influence of Zn-oxidation processes, not that of processes involving VO_n complexes, whose presence could as well influence the equilibrium concentration of vacancies in both Zn-diffusion and self-diffusion measurements.

The question remains, however, open though it seems demonstrated that the chemical nature of the diffusing impurity and the chemical reactivity of point defects might have a crucial, but obvious, role on the outcomes of impurity-diffusion measurements.

Chapter 4

4. Defects in group IV carbides

4.1. Structure and defects of carbides: generalities

Experimental evidence of point (and extended) defects in group IV carbides is only available for silicon carbide single crystals, epitaxial layers and microcrystalline- or amorphous-SiC films, while only results of theoretical evaluations are known for GeC [196].

Being SiC a material stable in more than 200 different polytypes, we expect to find in it a variety of native point defects, point defect complexes and extended defects, whose structure and properties would depend on the polytype nature, making at least critical their study.

The analysis here will be limited to few of its polytypes, the cubic (zinc blende) 3C-SiC and the hexagonal 4H and 6H ones, to which major attention in literature was so far deserved.

Fig. 37 [197] displays the configuration of a carbon atom (red in the figure) in a cubic (a) or hexagonal configuration (b). It should be noted that they present the same configuration of the nearest neighbor Si atoms, but a different configuration of the second (carbon) neighbors (black spheres), which lie on the top of the silicon atoms in the cubic site but are rotated by 60° in the case of the hexagonal site.

In the 3C-SiC polytype only cubic sites exist (see fig.37 left) and only hexagonal sites are present in the 2H polytype (see fig.37 right). Alternate cubic and hexagonal sites are instead present in the 4H polytype.

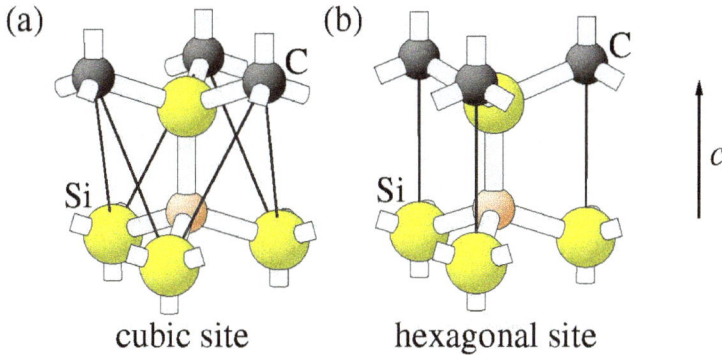

Fig 37 Local coordination of Si and C atoms in a cubic site (a) and in an hexagonal site (b) of SiC. After A. Mattausch [197]. With kind permission from Alexander Matthausch.

Independently of the polytype structure, but with different local configurations, native point defects should be found in both SiC sublattices. In fact, both Si-vacancies V_{Si} (see Fig.38) and C- vacancies V_C (Fig.39) are present in SiC, where, due to the different electronegativities of C and Si, the silicon vacancy has the nature of a cationic vacancy and the carbon vacancy that of an anionic vacancy, close to the case of the III-V compound semiconductors (GaAs, InAs).

The defect concentration should be influenced by the SiC non-stoichiometry, which is expected, from trivial thermodynamic considerations, to depend on the growth- and further annealing -processes, and on the nature of the SiC polytype involved.

Preferential sublimation of silicon

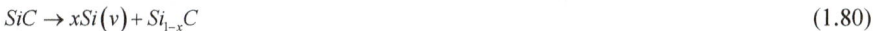

$$SiC \rightarrow xSi(v) + Si_{1-x}C \tag{1.80}$$

or Si-oxidation

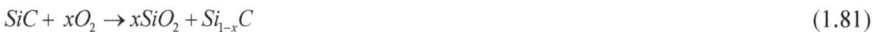

$$SiC + xO_2 \rightarrow xSiO_2 + Si_{1-x}C \tag{1.81}$$

would lead to substoichiometric, silicon deficient (i.e. carbon-rich) SiC phases[52].

---

[52] All these processes are not equilibrium ones.

Analogously, preferential oxidation of carbon

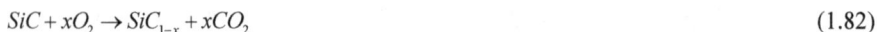

$$SiC + xO_2 \rightarrow SiC_{1-x} + xCO_2 \qquad (1.82)$$

could lead, instead, to a sub-stoichiometric, carbon deficient (i.e. Si-rich) SiC phase.

All these processes could occur simultaneously, ruled by the thermodynamics (and kinetics) of the relative processes, with the potential formation of a sub-stoichiometric, Si-and C-deficient, phases.

In the case of a silicon excess (i.e. carbon deficit) in the SiC phase, one expects, as an example, the formation of an excess of silicon interstitials, assuming that the silicon in excess would sit in interstitial sites

$$Si \rightarrow Si_i \qquad (1.83)$$

over its equilibrium concentration, or in new silicon sites with the simultaneous formation of carbon vacancies

$$Si \rightarrow Si_{Si} + V_C \qquad (1.84)$$

and a consequent shift of silicon vacancies and carbon self-interstitials from their equilibrium concentrations in stoichiometric SiC.

This issue was investigated by Birnie and Kingery [198], who determined the amounts of excess carbon and silicon dissolved in 3C-SiC and 6H-SiC samples at 2400 °C, by heating a two phase mixture consisting of carbon and SiC powder and silicon and SiC powder, respectively.

It was possible to demonstrate that the single-phase materials, extracted from a two-phase matrix from where the excess of carbon or silicon was suitably removed[53], presented densities almost coinciding with the theoretical densities and that the deviations from the stoichiometry determined by chemical analysis remained within the measurement errors. The conclusion was that the solubility of C and Si in SiC is negligibly small and that both polytipes are virtually stoichiometric. This hypothesis does not hold for epitaxially-grown SiC samples [201-202], where a C-excess or C-deficit would depend on growth conditions.

---

[53] Excess carbon was removed by oxidation and excess silicon by HF etch on finely crushed powder.

Therefore, at least up to 2400 °C, stoichiometric deviation could be neglected in the experimental and theoretical analysis of defects in SiC ingots, allowing to make the hypothesis that Schottky

$$Si_{Si}^{Sisl} \rightleftharpoons V_{Si}^{Sisl} + Si_{surf} \qquad (1.85)$$

$$C_{C}^{Csl} \rightleftharpoons V_{C}^{Csl} + C_{surf} \qquad (1.86)$$

$$Si_{surf} \rightleftharpoons Si_{i} \qquad (1.87)$$

$$C_{surf} \rightleftharpoons C_{i} \qquad (1.88)$$

and Frenkel processes

$$V_{Si}^{Sisl} + I_{Si}^{Sisl} \rightleftharpoons S_{Si} \qquad (1.89)$$

$$V_{C}^{Csl} + I_{C}^{Csl} \rightleftharpoons S_{C} \qquad (1.90)$$

would be simultaneously active in the silicon and carbon sublattices[54] of SiC, with the formation of Frenkel pairs.

Given the fact that we expect the formation of different self-interstitial types, like in the case of elemental semiconductors, multiple equilibria under eq.1.87, 1.88, 1.89 and 1.90 would occur, involving all stable self-interstitials.

Eventually, different from the elemental semiconductors case, also the presence of silicon antisites Si_{C} and of carbon antisites C_{Si} is expected, consisting of silicon atoms hosted in the C-sublattice and of carbon atoms hosted in the silicon sublattice.

In analogy with the case of elemental semiconductors, the experimental defect evidence comes from self-diffusion measurements [199-202], while their spectroscopic signatures are obtained using EPR-, photoluminescence (PL)-, IR-Raman-, Positron Annealing-Spectroscopy (PAS)[55], Optically detected Magnetic resonance (ODMR) and DLTS (Deep level Transient spectroscopy) measurements [203].

Positron annealing spectroscopy (PAS) in its various configurations, as a first example, has been used to identify the nature of the radiation-induced defects in 3C, 6H and 4H SiC from the positron lifetime spectra recorded after the irradiation and during temperature annealing cycles [204]. PAS results indicate that the dominant vacancy-type

---

[54] We use the apex Si sl or C sl to indicate the Si and C sublattices.
[55] It should be mentioned here again that the intrisic advantage of a PAS signature, measured as an increase of the positron lifetime, is the identification of a vacancy-type of defect.

of defect created by 2MeV electron irradiation at room temperature of a nitrogen-doped 3C-SiC sample is a Si-vacancy [205], whereas the C-vacancy has not been detected due to the weak localization of positrons at V_C. The carbon vacancy has been, instead, found in 4H-and 6H-SiC.

EPR experiments succeeded in the identification of the negative silicon- and carbon vacancies and gave the signatures of the carbon vacancy and of silicon antisite defects [203].

Photoluminescence (PL) spectroscopy and Optically Detected Magnetic Resonance (ODMR) give further information on vacancy-related defects [206], but the identification of the nature of defects from experimental data is, however, and in general, a very difficult task, also because isolated defects created during irradiation might interact and form different defect centres, making often tentative their atomistic identification [207].

Additional experimental difficulties arise from the presence of morphological defects (defect clusters, stacking faults, second phases), which are formed during the irradiation with high energy particles and may remain present after thermal annealing [208].

4.2   Theoretical and experimental evidences

Most of the information available on the properties of isolated point defects (electronic structure, formation and migration energies) in SiC arises from theoretical studies [207, 209-216].

Ab initio calculations of the formation energies of silicon- and carbon vacancies (see Table 10) show a large variance of their thermodynamic stabilities, but suggest that the carbon vacancy is the dominant defect in 3C-SiC [215]. They suggest also the silicon vacancy behaves as a metastable defect [216] and converts in a carbon vacancy-carbon antisite pair [203] in p-type SiC

$$V_{Si} \rightarrow V_C C_{Si} \qquad\qquad (1.91)$$

while is thermodynamically stable in n-type material 197,203, 217-220].

The carbon vacancy is instead stable in all doping conditions [217].

The different thermodynamic stability of carbon and silicon vacancies derives on the different nature of the carbon and silicon dangling bonds.

In the case of the silicon vacancy, the carbon dangling bonds are, in fact, strongly localized at the carbon atoms and the equilibrium configuration of the defect, as well the associated formation energy, results from the strong Coulomb interaction among localized electrons which minimizes the energy gain by bond formation.

In the case of carbon vacancy, the silicon dangling bonds are, instead, extended, allowing their overlap and leading to bond reconstruction and Jan Teller distorsion [203] with a resultant lower energy of formation.

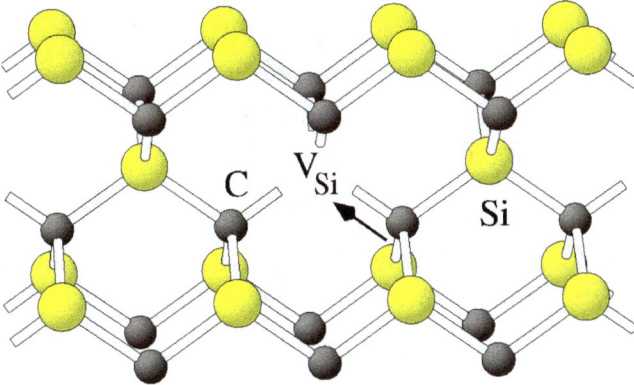

Fig. 38 The structure of a silicon vacancy in the 3C-SiC polytype. After A. Mattausch [197]. With kind permission from Alexander Matthausch.

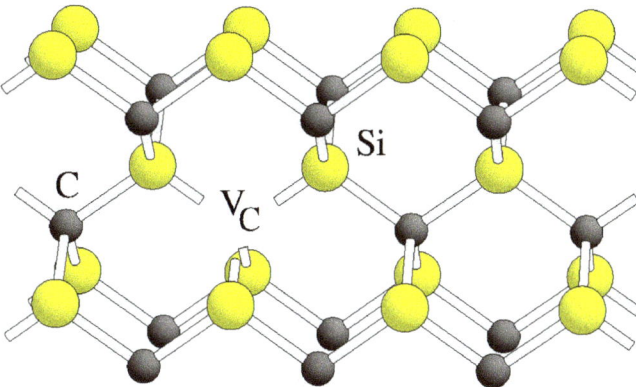

Fig. 39 The structure of a carbon vacancy in the 3C-SiC polytype. After A. Mattausch [197]. With kind permission from Alexander Matthausch.

Table 10. Calculated formation energies (in eV) of neutral carbon- and silicon vacancies in 3C-SiC [220].

V_c	V_{si}	References
5.6	7.6	Wang [209]
4.01	8.74	Zywietz [212]
3.78	8.34	Bockstedte [213]
3.84	8.78	Bernardini [215]
3.74	8.37	Torpo [207]
	9.70	Bruneval [216]

Table 11. Calculated formation energies (in eV) of neutral carbon- and silicon- antisites and of the carbon vacancy-carbon antisite complex in 3C-SiC [220].

C_{Si}	Si_C	V_C-C_{Si}	References
4.10	3.50	7.29	Mattausch (2005)
4.15	3.66	7.24	Bernardini (2004)
4.3	4.0		Torpo (1998)
		7.45	Bruneval (2011)

Antisite formation in SiC is favored by the close interconnection of Si_4C and C_4Si tetrahedral units, which makes possible a direct position exchange

$$C_C + Si_{Si} \leftrightarrow C_{Si} + Si_C \tag{1.92}$$

Given their calculated formation energies (see Table 11), comparable to the formation energy of carbon vacancy, antisites are the most common defects in SiC [215].

As shown before, the formation of an antisite could be mediated by a vacancy.

In this case the exchange process could be formally written as

$$V_{Si} \rightleftharpoons V_C C_{Si} \tag{1.93}$$

that could be better viewed using the notation introduced in Section 2.4.5

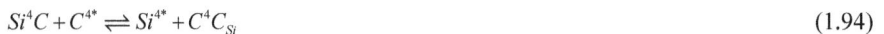

$$Si^4C + C^{4*} \rightleftharpoons Si^{4*} + C^4 C_{Si} \tag{1.94}$$

which describes the transfer of a carbon atom from the center of a tetrahedron Si^4C having four Si atoms at the corners to the empty center of a neighbor silicon vacancy C^{4*}, with the contemporaneous formation of a close pair of a carbon vacancy Si^{4+} and of a carbon antisite C^4C_{Si}. This close pair behaves as a thermodynamically stable defect complex (see Fig.40).

One can see, in fact, in Table 11, that the calculated formation energy of the neutral complex $V_C.C_{Si}$ (≈ 7.3 eV) is less than the formation energy of the silicon vacancy (≈ 8.5eV) enabling an energy gain of more than 1 eV associated to its formation. According to Mattausch [197] the energy gain raises to 4 eV for a Fermi level at the valence band. This makes of the silicon vacancy a metastable defect, but since the activation energy for the process of $V_C.C_S$ formation is rather high (between 1.9 and 2.7 eV [197] the process under eq. 1.93 and 1.94 should occur only at high temperatures.

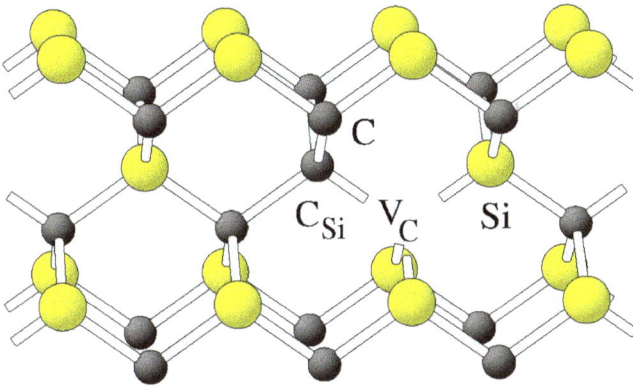

Fig. 40 The carbon vacancy-silicon antisite complex. After A. Mattausch [197]. With kind permission from Alexander Matthausch.

A process mediated by a carbon vacancy V_C, leading to the formation of a silicon antisite Si_C and a silicon vacancy V_{Si}, while formally possible

$$V_C \rightleftharpoons Si_C + V_{Si} \qquad (1.95)$$

is, instead, inhibited by the thermodynamic stability of the carbon vacancy.

Since SiC shares with diamond hardness and lattice stiffness, we expect, like in diamond, large formation energies of silicon- and carbon interstitials [221]. Table 12 displays their calculated values, which are, in fact, quite higher than those of the carbon vacancy and of the silicon and carbon antisites.

Table 12. Calculated formation energies (in eV) of neutral carbon and silicon self-interstitials in 3C-SiC [220].

I_C split <100>	I_C split <110>	I_{Si}split<100>	I_{Si}<T>	References
6.64	6.84	7.07	5.67	[214]
6.64	6.84	7.07	5.67	[214]
6.73	7.31	8.59		[213]
	7.3			[215]
7.2	6.7			[210]

While for individual defects a reasonable agreement is observed between *ab initio* theory predictions and experimental evidence, theory is still unable, in spite of the progress made in the last recent years [220], to work satisfactorily with the more sophisticated problems of defect aggregation and defect complexes formation in SiC, which are the most frequently observed conditions occurring in practice.

Here, according to Bockstedte [203] the investigations *are guided by chemical intuition and physical experience*[56] especially considering that a material like SiC is by itself dramatically prone also to local phase transformations, with a consequent increase of number of different radiation damage defects and, potentially, large local non-stoichiometry depending on the irradiation conditions.

This is, as an example, the case for homoepitaxial[57] SiC samples studied by the present author [222-223], before and after irradiation with 8.2 MeV electrons.

The measurements were carried-out using PL spectroscopy in the 1-4 eV region, since most defects were expected to exhibit optical signatures in this energy range.

One can see (see Fig. 41) that the broad PL band of the virgin sample could be deconvoluted in three bands peaked at 1.9, 2.4 and 2.6 eV, of which that at 1.9 eV falls in a region of intrinsic defect centres involving carbon vacancies [208] (in close agreement with the Coulson [5] hypothesis that the emission around to 2 eV is due to a carbon vacancy in diamond) and that at 2.6 eV could be assigned to a B_{Si}-C pair [224].

It is interesting to observe that after irradiation (Fig. 42) only an emission at 2.4 eV is present, with a tail extending toward low energies, while after annealing at 450 °C the pristine situation of the virgin sample is recovered.

---

[56] Original sentence in the paper of Bockstedte.
[57] Over single crystal 4H SiC samples.

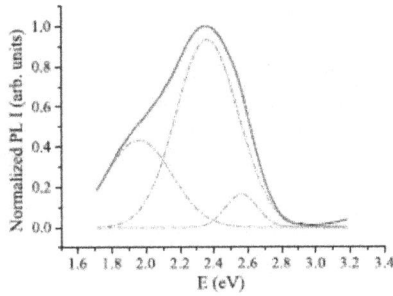

Fig. 41 PL spectrum of a virgin SiC sample. After A. LeDonne et al [222]. Reproduced with permission from Elsevier, License number 3994250984699, Nov.22, 2016.

Since radiation damage, in absence of annealing, should mostly involve the formation of point defects (vacancies and interstitials) and impurity complexes, one could qualitatively conclude from these experiments that the band at 2.4 eV might be associated to point defects and to their complexes with impurities, and that thermal annealing of the irradiated samples recovers the original situation of the virgin sample, via thermally-enhanced reactions of native defects with dopant impurities, B in this case.

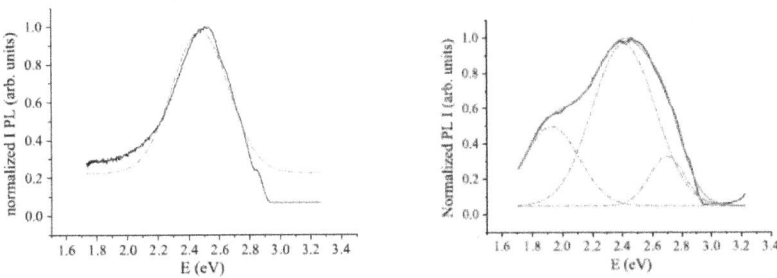

Fig. 42. Pl spectrum of an irradiated SiC sample (left) after annealing at 450 °C (right). After A. LeDonne et al [222]. Reproduced with permission from Elsevier, License number 3994250984699, Nov.22, 2016.

4.3 Selfdiffusion

Self-diffusion studies initiated in the sixties of the last century on samples of dubious quality, since at that time the Lely process was still a delicate and dedicated laboratory technique. The first measurements recorded in literature were carried out by Ghoshtagore and Coble [200] on Al-doped and N-doped, formally single crystalline samples.

They succeeded in demonstrating (see Table 13) that the carbon diffusivity in both p-type and n-type samples is accounted by an exponential dependence on temperature. However, while the activation energy for the p-type samples is reasonably close to an expected dependence on vacancy migration, the extremely high (13 eV) activation energy of n-type samples is incompatible with a vacancy mediated mechanism.

The results of self-diffusion measurements carried out few years later by Hong *et al* [199,225-227] on hexagonal (single crystal) and cubic (polycrystalline) SiC, using ^{14}C and ^{30}Si, fit well with an exponential dependence on temperature of the self-diffusion coefficients of both carbon and silicon, and show small, but non negligible differences in the activation energies of 4H and 3C samples. Overall, see Fig. 43 (upper curves) and table 13, results are compatible with a defect mediated transport.

The main difference between the absolute values of Si- and C- diffusivities depends on a large difference of their relative prefactors (see Table 13, reference 227) not easily explainable since the activation energies are very close together, and consistent with the values reported in later works.

The same measurements also showed that the diffusion coefficients are sensible to doping, and thus, to the charge of the defects mediating the transport. The original conclusion of this work was that in both cases one relies on a vacancy mediated transport.

Also later work [198] carried out on single crystalline samples showed similar activation energies for C and Si diffusion, anticipating the results and conclusions of Rüschenschmidt *et al* [202], who carried out self-diffusion experiments on high purity, CVD-grown epitaxial layers on 4H SiC.

They demonstrated (see Fig. 43 and Table 13) not only that the carbon and silicon diffusivity data overlap on an Arrhenius plot in the 2000-2200 °C range, but that this exponential dependence coincides also with that of the defect-mediated diffusivity of B-dopant in SiC, with an activation energy of 7.6 ± 1 eV.

The authors, therefore, suppose the existence of a unique defect species which mediates the C and Si migration in both sublattices, without reaching, however, a univocal conclusion common to C, B and Si, but excluding *a priori* a direct role of carbon and silicon vacancies. This assumption is based on the reasonable unlikelihood of silicon transport assisted by carbon vacancies and of a carbon transport mediated by silicon vacancies, independently of the difference in their transport capacities and defect formation energies (see Tables 10, 13 and 14) that would favor carbon vacancies or carbon self-interstitials, depending on the non-stoichiometry.

Eventually, the C-split interstitial is considered an appropriate vehicle for both C- and B-diffusion, but the case of silicon remains unsolved.

Table. 13 Self-diffusion data on 3C- and 4H-SiC.

Diffusing species	Sample	Temperature range (°C)	Self-diffusion coefficient (cm²sec⁻¹)	Ref.
^{13}C	4H-SiC (intrinsic)	2100-2350	$D_C = 8.4.10^2 \exp - \dfrac{8.50eV}{kT}$	201
^{30}Si and ^{13}C	4H-SiC (CVD grown)	1700-2200	$D = 4.8 \pm^{573}_{4.7} \exp - \dfrac{7.6 \pm 1eV}{kT}$	202
^{14}C and ^{30}Si	4H-SiC s.c	1850-2300	$D_C = 8.62(\pm 2.01)10^5 \exp - \dfrac{7.41 \mp 0.05eV}{kT}$ $D_{Si} = 1.54(\pm 0.78)10^5 \exp - \dfrac{8.10 \pm 0.10eV}{kT}$	227
^{14}C and ^{30}Si	β-SiC (3C-SiC) poly	1850-2300	$D_C = 1.2(\pm 1.83)10^8 \exp - \dfrac{8.72 \pm 0.14}{kT}$ $D_{Si} = (8.36 \pm 1.99)10^7 \exp - \dfrac{9.45 \pm 0.05eV}{kT}$	227
^{14}C	6H p-type 600ppm Al	1853 -2060	$D_C = 3.10^2 \exp - \dfrac{6.114 \pm 0.719eV}{kT}$	200
^{14}C	6H n-type 100ppm N	1977-2088	$D_C = 2.10^{17} \exp - \dfrac{13.11 \pm 2.107eV}{kT}$	200

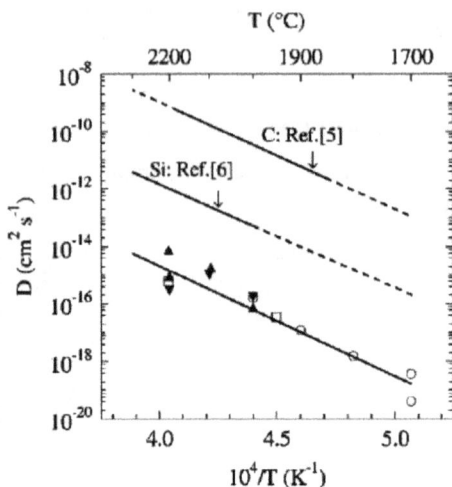

Fig. 43 Self-diffusion coefficients of C and Si in 4H-SiC. Upright triangles for Si-, reverse triangles for C-diffusivities, while open symbols refer to their diffusivities deduced from B-diffusion. After K.Rüschenschmidt et al [202]. The upper curves display the Hong's results concerning Si and C self-diffusion. Reproduced with permission from AIP Publishing LLC, License number 4019400977368, License date Dec 31, 2016.

This conclusion agrees with that deduced from the measurement of C self-diffusion in intrinsic 4H-SiC, carried out by Linnarsson et al [201] who showed (see Fig. 44) an excellent Arrhenius behavior of their data, in the 2100-2350 °C range, which leads to an activation energy of 8.50 eV, coincident, within the experimental errors, with that of Rüschenschmidt et al [202] and in the range of the Hong's ones [199, 225-227].

On the base of the calculated activation energies for C self-diffusion in C-rich and Si-rich SiC [219] (see Table 14), Linnarsson et al [201] concluded that in carbon-rich conditions the C self-interstitial migration should be favored, while in silicon-rich conditions the C-vacancy and the C-interstitial mediated diffusion should be equally viable.

These conclusions fit well with the role assigned to C self-interstitials assigned by Rüschenschmidt et al [202], leaving however again unsolved the mechanism of Si diffusion.

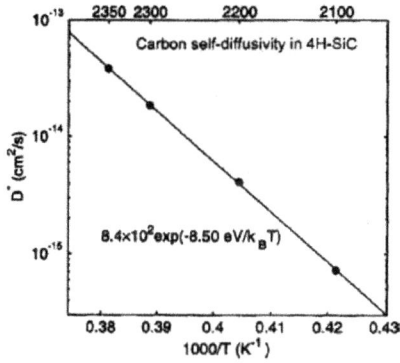

Fig. 44 Temperature dependence of carbon self-diffusion in intrinsic 4H-SiC. After M. K. Linnarsson et al [201]. Reproduced with permission from AIP Publishing LLC, License number 4019411334814, License date Dec 31, 2016.

Table 14 Activation energies for C self-diffusion (eV) for vacancy and self-interstitial mediated processes. After M. Bockstedte et al [213].

	Si-rich	C-rich
V_C^{2+}	6.6-9	7.2-9.6
V_C^0	7.3	7.9
$I_{split<100>}^{2+}$	6.7-7.8	6.1-7.2
$I_{split<100>}^{1+}$	7.2-7.9	6.6-7.3
$I_{split<100>}^{0}$	7.2	6.6
$I_{split<100>}^{-1}$	6.7 -7.3	6.1-6.7

The possibility remains of a direct exchange process within neighbor antisites positions, with migration paths occurring along both the C and Si lattices, favored by the low formation energies of both the C-and Si-antisites.

Though the good fit of the Linnarson *et al* [201] and Rüschenschmidt *et al* [202] results seem to show that the structural and chemical quality of the SiC samples plays a key role, as trivially could be expected, on the homogeneity and quality of the results, still the apparent need of a unique species mediating the Si, C and B transport in SiC remains an unknown, difficult to be clarified, but exiting to be eventually understood.

Conclusions

After several decades of experimental and theoretical research, defects in semiconductors represent a still exciting challenge provided experiments would be carried-out with major accomplishments of the chemical nature of the systems under study, as brilliantly shown by Bracht *et al* in a recent work.

Nanostructured semiconductors represent the immediate present and the future of these research activities, on which a preliminary analysis should be soon devoted.

Acknowledgments

The author is strongly indebted to: Prof. D.P. Birnie III, Department of Materials Science and Engineering, Rutgers University; Prof. H. Bracht, Institute of Materials Physics, University of Muenster; Prof. S. Clark, Physics Department, Durham University; Dr. G. Kissinger, Leibniz-Institut für innovative Mikroelektronik, Frankfurt (Oder); Prof. C. Londos, Physics Department, University of Athens; Dr. Alexander Mattausch and Prof. Michel Bockstedte, Erlangen-Nuremberg University; Prof. B. Svensson, Physics Department, University of Oslo; Prof. Hichiro Yonenaga, Institute for Materials Research, Tohoku University   and to Mette Fjelltveit Rye-Larsen, Physics Department, University of Oslo for suggestions, copies of papers and figures used in this book.

The author would also like to warmly thank Sarah for having drawn several figures for this book.

References

[1]   W. Schottky (1935) Über den Mechanismus der Ionenbewegung in festen Elektrolyten Z. Phys. Chem B29 335-355; https://doi.org/10.1515/zpch-1935-0134

[2]   C. Wagner (1936) Über die Natur der Fehlordnungschienigungen in Silber Bromid Z.Phys.Chem. B32 113-116]

[3]   T. Okada, T. Litaka, T. Yagi, K. Aoki (2014) Electrical conductivity of ice VII Nature, Scientific reports 4 5778 DOI:10.1038/srep05778, https://doi.org/10.1038/srep05778

[4]   S. Geller, J. R. Akridge, S.A. Wilber (1979) Crystal structure and conductivity of the solid electrolyte a-RbCu4Cl3I2 Phys. Rev. B 19 5396 – 5402. https://doi.org/10.1103/PhysRevB.19.5396

[5]   C. A. Coulson, M.J. Kearsley (1957) Colour centres in irradiated diamond Proc. R. Soc A 241:433-454 https://doi.org/10.1098/rspa.1957.0138

[6]   G.D Watkins (1977) Lattice vacancies and Interstitials in silicon Chin. J.Phys. 15 92-10

[7]   A. Chroneos, C.A.Londos, En. Sgourou (2014) Strategies to suppress A-center formation in silicon and germanium from a mass action analysis viewpoint physics.karazin.ua/doc/v_20_2014/pdf/1113/87-90

[8]   S. Pizzini, (2015) Physical chemistry of semiconductor materials and Processes, J. Wiley&Sons https://doi.org/10.1002/9781118514610

[9]   J. Bernholc, A. Antonelli, T. M. DelSole, Y. Bar-Yam, S.T. Pantelides (1988) Mechanism of self-diffusion in diamond Phys. Rev.Lett. 61 2689 https://doi.org/10.1103/PhysRevLett.61.2689

[10]   H. Kwart, K. King (1977) d-Orbitals in the Chemistry of Silicon, Phosphorus and Sulfur Springer-Verlag, Berlin, Heidelberg

[11]   J. Narayan, A. Bhaumik (2015) Novel phase of carbon, ferromagnetism, and conversion into diamond J. Appl. Phys. 118, 215303 https://doi.org/10.1063/1.4936595

[12]   A. Kailer, Y. G. Gogotsi, and K. G. Nickel (1997) Phase transformations of silicon caused by contact loading J.Appl.Phys. 81, 3057-3064 https://doi.org/10.1063/1.364340

[13]   W. Jank, J. Hafner (1988) The electronic structure of liquid germanium Europhys.Lett.7 623-628] https://doi.org/10.1209/0295-5075/7/7/009

[14]   A. Jayaraman, W. Klement, Jr, G. C. Kennedy (1963) Melting and Polymorphism at High Pressures in Some Group IV Elements and III-V Compounds with the Diamond/Zincblende Structure Phys. Rev. 130, 540 -547 https://doi.org/10.1103/PhysRev.130.540

[15]   B. H. Cheong and K. J. Chang (1991) First principle study of the structural phase transition in Sn Phys.Rev.B 44 4103-4108

[16]   S-H. Na, C-H. Park (2010) First principle study of the structural phase transition in Sn J. Korean Phys.Soc. 56 494-497 https://doi.org/10.3938/jkps.56.494

[17]   J. Fleeman, G. J. Dienes (1955) Effect of Reactor Irradiation on the White-to-Grey Tin Transformation J.Appl. Phys. 26 652 - 654 https://doi.org/10.1063/1.1722064

[18]   K. Yada, K. Torigoe (1973) Electron-microscopic evidence of transformation-induced lattice defects in grey tin J. Maters.Sci. 8, 297-298 https://doi.org/10.1007/BF00550683

[19]   E.S. Hedges and J.Y. Higgs (1952) Preparation of Grey Tin Nature 169 621-622

[20]   G.A. Busch, R.Kern (1960) Semiconducting properties of grey tin. Solid State Physics 11 1-39 https://doi.org/10.1016/S0081-1947(08)60166-6

[21]   S. Groves and W. Paul (1963), Band Structure of grey tin Phys.Rev.Lett. 11 194 https://doi.org/10.1103/PhysRevLett.11.194

[22]   R. J. Naumann (2008) Introduction to the Physics and Chemistry of Materials, CRC Press

23]   I. Stich, R. Car, M. Parrinello (1989) Bonding and disorder in liquid silicon Phy.Rev.Lett.63 2240-2243 https://doi.org/10.1103/PhysRevLett.63.2240

[24]   J. T. Okada, P-H. Sit, Y. Watanabe et al. (2012) Persistence of covalent bonding in liquid silicon probed by inelastic x-ray scattering Phys. Rev.Lett. 108, 067402 https://doi.org/10.1103/PhysRevLett.108.067402

[25]   M. Davidovic, M Stojic and D Jovic (1983) The study of liquid germanium structure J. Phys. C: Solid State Physics 16 2053-2058 https://doi.org/10.1088/0022-3719/16/11/008

[26]   W. Jank and J. Hafner (1988) The Electronic Structure of Liquid Germanium, EPL (Europhysics Letters) 7 623-628 https://doi.org/10.1209/0295-5075/7/7/009

[27]   M. van Thiel and F. H. Ree (1993) High-pressure liquid-liquid phase change in carbon Phys. Rev. B 48 3591-3599 https://doi.org/10.1103/PhysRevB.48.3591

[28]   L. M. Ghiringhelli and E. J. Meijer, (2010) Liquid Carbon: Freezing Line and Structure Near Freezing in L. Colombo and A. Fasolino (Eds.), Computer-Based Modeling of Novel Carbon Systems and Their Properties, Carbon Materials: Chemistry and Physics 3, DOI 10.1007/978-1-4020-9718-8 1, pp.1-36 Springer Science+Business Media B.V.

[29] R. W. Olesinski, G. J. Abbaschian, (1984) The Si-Sn (Silicon-Tin) system, Journal of Phase Equilibria 5(3) 273-276 https://doi.org/10.1007/bf02868552

[30]R. W. Olesinski, G. J. Abbaschian (1984) The Ge-Sn (Germanium-Tin) system Bulletin of Alloy Phase Diagrams 5(3) 265-271

[31] T. Soma (1979) The electronic theory of Si-Ge solid solution Phys. Stat. Solid. (b), 95 427-431 https://doi.org/10.1002/pssb.2220950212

[32] I. Yonenaga (2015) Growth and Characterization of silicon-germanium alloys in Silicon, Germanium and their alloys G. Kissinger and S.Pizzini Edts. CRC press

[33] R. Pandey, M. Rerat, Cl. Darrigan, M. Causà (2000) A theoretical study of stability, electronic, and optical properties of GeC and SnC J.Appl.Phys.88 6462-6466 https://doi.org/10.1063/1.1287225

[34] R.I. Scace, G.A.Slack (1959) Solubility of carbon in silicon and germanium J.Chem. Physics 30 1551-1555 https://doi.org/10.1063/1.1730236

[35] E.E. Haller, W. L. Hansen, P. Luke, R. McMurray and B. Jarrett (1982) Carbon in high purity germanium IEEE Trans.Nucl. Sci. 29 745-750 https://doi.org/10.1109/TNS.1982.4335949

[36] D. Both, K. Voss (1981) The optical and structural propertiesof CVD germanium carbide J.Phys. (France), Coll. 42(C4) 1033-1036

[37] J. T. Herrold, V.L. Dalal (2000) Growth and properties of microcrystalline germanium carbide allyos grown using electron cyclotron resonance plasma processing J.Non-Cryst.Solids 270 255-259 https://doi.org/10.1016/S0022-3093(00)00091-0

[38] M. A. Fraga, R. S. Pessoa, H. S. Maciel, M. Massi (2011) Recent Developments on Silicon Carbide Thin Films for Piezoresistive Sensors Applications in Silicon Carbide – Materials, Processing and Applications in Electronic Devices, M. MukherjeeEdt. pp.369- 388 InTech Publ.

[39] L. Liu, W. Tang, B-X. Zheng and H.-X Zhang, (2011) Fabrication and Characterization of SiC Thin Films Proceedings of the 2011 6th IEEE International Conference on Nano/Micro Engineered and Molecular Systems February 20-23, 2011, pp.146-149 Kaohsiung, Taiwan

[40] D. A. Anderson , W. E. Spear (1977) Electrical and optical properties of amorphous silicon carbide, silicon nitride and germanium carbide prepared by the glow discharge technique Phil.Magaz. 35 1-16 https://doi.org/10.1080/14786437708235967

[41]   G. Leal, T. M. Bastos Campos, A. S. da Silva Sobrinho, R. S. Pessoa, H.S. Maciel, M. Massi (2014)Characterization of SiC thin films deposited by HiPIMS Mat. Res.17, Print version ISSN 1516-1439

[42]   R. Yakimova and E. Janzen (2000) Current Status and Advances in the Growth of SiC Diamond and Related Materials 9 432–438 https://doi.org/10.1016/S0925-9635(99)00219-8

[43]   G. Dhanaraj, X.R. Huang, M. Dudley, V. Prasad, R.-H. Ma, (2003) Silicon Carbide Crystals — Part I: Growth and Characterization in Crystal Growth technology, K.Byrappa, W.Michaeli, H.Waarlimont, E.Weber (Eds) William Andrew Inc.

[44]   M. Pons, R. Madar, T. Billon (2004) Principles and limitations of numerical solutions of SiC boule growth by sublimation in Silicon Carbide. Recent mayor advances W.J. Choyke, H.Marsunami, G.Pensl (Eds) ISBN 3-340-40458-9 pp-121-136Springer- Verlag Berlin, Heidelberg New York

[45]   W. L. Jolly, D. Latimer (1952) The Equilibrium Ge(s) + GeO,(s) = 2GeO(g). The Heat of Formation of Germanic Oxide J.Am. Chem Soc.74 5757-5758 https://doi.org/10.1021/ja01142a056

[46]   R.F. Davis, G. Kelner, M. Shur, J.W. Palmour, J.A.Edmond (1991) Thin Film Deposition and Microelectronic and Optoelectronic Device Fabrication and Characterization in Monocrystalline Alpha and Beta Silicon Carbide Proceed. IEEE 79 677-701 https://doi.org/10.1109/5.90132

[47]   D. Both, K. Voss (1981) The optical and structural propertiesof CVD germanium carbide J.Phys. (France), Coll. 42(C4) 1033-1036

[48]   A. Mahmood, L. Enrique Sansores (2005) Band structure and bulk modulus calculations of germanium carbide J.Mat. Res. 20 1101-1106 https://doi.org/10.1557/JMR.2005.0172

[49]   A.F. Sankey, A.A. Demov, W.T. Petuskey and P.F. McMillan (1993) Energetics and electronic structure of the hypothetical cubic zincblende form of GeC. Modell. Simul. Mater. Sci. Eng. 1 741 https://doi.org/10.1088/0965-0393/1/5/014

[50]   A.H. Lettington, C.J.H. Wort and B.C. Monachan (1989) Developments and IR applications of GeC thin films. Proc. SPIE 1112 156 https://doi.org/10.1117/12.960774

[51]   C. A. Madu (2014) Electronic and Structural Properties of the Silicon and Germanium Carbides Res. J. Physical Sci. 2 1-5

[52]   M. Lanoo and J. Burgoin (1981) Point Defects in semiconductors Springer Verlag

[53]   W. Schröter (Edt) (1991) Electronic structure and properties of semiconductors in Materials Science and Technology, Vol.4 pp 321-378 Wiley-VCH Verlag Gmbh Wienheim

[54]   P. Pichler (2004) Intrinsic Point Defects, Impurities, and Their Diffusion in Silicon Springer., Wien New York https://doi.org/10.1007/978-3-7091-0597-9_2

[55]   C. Claeys, E. Simoen (2007) Germanium -based technologies: from Materials to Devices. Elsevier

[56]   H. Meher (2007) Diffusion in Solids: Fundamentals, Methods, Materials, Technology and Engineering Springer Science & Business Media https://doi.org/10.1007/978-3-540-71488-0

[57]   A. Drabold, S. Estreicher, (Eds.) (2007) Theory of Defects in Semiconductors, Springer Verlag, Berlin Heidelberg

[58]   G. Kissinger, S. Pizzini (Edt) (2015) Silicon, Germanium and Their alloys: Growth, Defects Impurities and nanocrystals , CRC Press, Taylor and Francis Group

[59]   L.Romano, V. Privitera, C. Jagadish (Eds) (2015) Defects in Semiconductors, 1st Edition, Elsevier

[60]   J. R. Weber, A. Janotti, and C. G. Van de Walle (2015) Defects in Germanium in Photonics and Electronics with Germanium, K. Wada and L.l C. Kimerling Edt., Wiley-VCH Verlag GmbH & Co.

[61]   Grey tin (alpha-Sn), impurities and defects in Subvolume A1b 'Group IV Elements, IV-IV and III-V Compounds. Part b - Electronic, Transport, Optical and Other Properties, Volume 41 'Semiconductors' Landolt-Börnstein - Group III Condensed Matter.

[62]   E. Kamiyama, K.Sueoka, I.Vanhellemont (2015) Vacancies in Si and Ge in Silicon, Germanium and their alloys G.Kissinger and S.Pizzini Edtrs pp.119-158 CRC Press

[63]   A.Carvalho and R.Jones, (2015) Self-Interstitials in silicon and germanium in Silicon,Germanium and their alloys G.Kissinger and S.Pizzini Edtrs. pp.87-118 CRC Press

[63] R. Car, P. J . Kelly, A. Oshiyama, S. T. Pantelides (1984) Microscopic Theory of Atomic Diffusion Mechanisms in Silicon Phys. Rev. Lett., 52 1814, https://doi.org/10.1103/PhysRevLett.52.1814

[64] G.B. Bachelet, G. Jacucci, R. Car, M. Parrinello (1987) Free energy of formation of lattice vacancies in silicon Proceed. 18thInt.Conf.Phys. Semicond. Stockholm (Sweden) O.Engstrom Edt. World Scientific, Singapore

[65] R. Car, P. E. Blöchl, and E. Smargiassi (1992) Ab-initio Molecular Dynamics of Semiconductor Defects, Material Science Forum 83-87 433-446 https://doi.org/10.4028/www.scientific.net/MSF.83-87.433

[66] E. Smargiassi and R. Car (1996) First-principles free-energy calculations on condensed-matter systems: Lattice vacancy in silicon. Phys. Rev. B53 9760-9763 https://doi.org/10.1103/PhysRevB.53.9760

[67] T. Nishimatsu, M. Sluiter, H. Mizuseki, Y. Kawazoe, Y. Sato, M. Miyata, M. Uehara (2003) Prediction of XPS spectra of silicon self-interstitials with the all-electron mixed-basis method Physica B: Condensed Matter 340–342 570–574 https://doi.org/10.1016/j.physb.2003.09.133

[68] S.A. Centoni, B. Sadigh, G.H. Gilmer, T.J. Lenosky, T. Diaz de la Rubia, C.B. Musgrave (2005) First-principle calculation on intrinsic defect volumes in silicon Phys.Rev B 72 195206 https://doi.org/10.1103/PhysRevB.72.195206

[69] R. Jones, A. Carvalho, J.P. Goss, P.R. Briddon (2008) The self-interstitial in silicon and germanium Mater. Science Eng.: B 159-160 112-116 https://doi.org/10.1016/j.mseb.2008.09.013

[70] J.R.Weber, A.Janotti, C. G. Van de Walle (2013) Dangling bonds and vacancies in germanium PhysRevB. 87 035203 https://doi.org/10.1103/PhysRevB.87.035203

[71] A. Zelferino, S. Salustro, J. Baima, V. Lacivita, R. Orlando, R Dovesi, (2016) The electronic states of the neutral vacancy in diamond: a quantum mechanical approach Theor.Chem. Acc. 74 135 (11 pages)

[72] J. Baima, A. Zelferino, P. Olivero, A. Erba and R. Dovesi (2016) Raman spectroscopic features of the neutral vacancy in diamond from ab initio quantum-mechanical calculations Phys. Chem. Chem. Phys., 18 1961 https://doi.org/10.1039/C5CP06672G

[73] J. Koike, D. M. Parkin, and T. E. Mitchell 1992) Displacement threshold energy for type Illa diamond Appl. Phys. Lett. 60 1450 https://doi.org/10.1063/1.107267

[74]   E. Monakkov, B.G.Svensson (2015) Point defect complexed in silicon in Silicon, Germanium, and their Alloys, G.Kissinger and S.Pizzini Edt. pp.255-288 CRCPress

[75]   S. Pizzini (2002) Chemistry and physics of defect interaction in Semiconductors in Defect interaction and clustering in semiconductors, S.Pizzini Ed. pp.1-68 Scitec Publ.Ltd CH-8707 Uetikon-Zurich/Switzerland ISBN 3-908450-65-9

[76]   J.Slotte, S.Kipellainen, F.Tuomisto, J.Raisanen, A.Nylandsted Larsen (2011) Direct observation of vacancies and its annealing in germanium Phys.Rev.B 83 235212 https://doi.org/10.1103/PhysRevB.83.235212

[77]   H. Alexander, H. Teichler (1991) Dislocations in Materials Science and Technology, Electronic structure and properties of semiconductors, W.Schroeter Edt.Vol.4 pp 321-378 Wiley-VCH Verlag Gmbh, Wienheim

[78]   S. G. Carter, E. R. Glaser, B. D. Weaver (2014) Room temperature Optically detected magnetic Resonance of silicon vacancies in SiC. APS March Meeting 2014, abstract #A36.011

[79]   H. Lim, S. Parj, H. Cheong, H.-M. Choi, Y.- C. Kim (2006) Photoluminescence of Natural Diamonds J. Korean Phys. Soc. 48 155-1559

[80]   J. Baima, A. Zelferino, P. Olivero, A. Erba and R. Dovesi (2016) Raman spectroscopic features of the neutral vacancy in diamond from ab initio quantum-mechanical calculations Phys. Chem. Chem. Phys., 18 1961-1968 https://doi.org/10.1039/C5CP06672G

[81]   J. P. Goss, B. J. Coomer, R. Jones, T. D. Shaw, P. R. Briddon and M. Rayson, S. Öberg (2001) Self-interstitial aggregation in diamond Phys. Rev.B 63 195208 https://doi.org/10.1103/PhysRevB.63.195208

[82]   J.W. Steeds, S.J. Charles, T.J.Davis A.Gilmore, J.Hayes, D.Pickard J.E. Butler (1999) Creation and mobility of self-interstitials in diamond by use of a transmission electron microscope and their subsequent study by photoluminescence microscopy Diamond and Related Materials 8 94-100 https://doi.org/10.1016/S0925-9635(98)00443-9

[83]   A. M. Stoneham (1978) The structure a nd motion of the selfinterstitial in diamond Solid State Electron. 21 1431-1433 https://doi.org/10.1016/0038-1101(78)90220-4

[84]   H. Bracht, E.E. Haller, R.Clark-Phelps (1998) Silicon self-Diffusion in Isotope Heterostructures Phys,Rev.Lett. 81 393-396 https://doi.org/10.1103/PhysRevLett.81.393

[85] H. Bracht (2015) Self-and Dopant diffusion in silicon, germanium and their alloys in Silicon, Germanium and their alloys, G. Kissinger and S.Pizzini Edt. pp 159-216 CRC Press

[86] V.V.Vo ronkov (1982) The mechanism of swirl defect formation in silicon J.Cryst. Growth 59 625-643 https://doi.org/10.1016/0022-0248(82)90386-4

[87] A. Seeger and P. Chick (1968) Diffusion mechanism and point defects in silicon and germanium phys. Status solidi 29 455-549 https://doi.org/10.1002/pssb.19680290202

[88] P. M. Fahey, P. B. Griffin, J. D. Plummer (1989) Point defects and dopant diffusion in silicon Rev. Modern Physics, 61 289-384 https://doi.org/10.1103/RevModPhys.61.289

[89] N.E.B. Cowern, S. Simdyankin, C. Ahn, N.S. Bennett, J.P. Goss, J.-M. Hartmann, A.Pafker, S.Hamm, J.Valentin, E.Napolitani, D.DeSalvasor, E.Bruno, S.Mirabella (2013) Extended Point defects in crystalline materials: Ge and Si Phys.Rev.Lett.110 15501 https://doi.org/10.1103/physrevlett.110.155501

[90] See supplemental material http:/link.aps.org/supplemental/10.1103/PhysRevLett,110,155501

[91] S. Clark (1994) Complex structures in tetrahedrally bonded semiconductors, Thesis Univ. Edinburgh

[92] F.P. Larkins, A.M. Stoneham (1971) Lattice distortion near vacancies in diamond and silicon J. Phys. C: Solid St. Phys. 4 154-163 https://doi.org/10.1088/0022-3719/4/2/003

[93] E.G. Song, E. Kim, Y.H. Lee (1993) Fully relaxed point defects in crystalline silicon Phys. Rev.B 48 1486-1489 https://doi.org/10.1103/PhysRevB.48.1486

[94] A. Antonelli, E. Kaxiras, D.J. Chadi (1998) Vacancy in silicon revisited: Structure and pressure effects Phys.Rev.Lett. 81 2088-2091 https://doi.org/10.1103/PhysRevLett.81.2088

[95] W.K. Leung, R.J. Needs, G. Rajagopal, S.Itoh, S.Ihara (1999) Calculation of silicon-self-interstitial Defect Phys.Rev.Lett. 82 2352 https://doi.org/10.1103/physrevlett.83.2351

[96] R.A. Brown, M.Maroudas, T.Sinno,(1994) Modelling point defect dynamics in the crystal growth of silicon J.Cryst.Growth 137 12-25 https://doi.org/10.1016/0022-0248(94)91240-8

[97]   V.Emsev, T.V.Mashovets, N.A,Vitovskii (1985) The spatial distribution of
       Frenkel pair components created in germanium and silicon under irradiation phys.
       status solidi (a) 90 523 - 530 https://doi.org/10.1002/pssa.2210900215

[98]   H. Zimmermann, H. Ryssel (1992) Gold and Pl atinum diffusion: the key for the
       understanding of intrinsic point defects in silicon Appl.Phys. A 55 121-134
       https://doi.org/10.1007/BF00334210

[99]   J. P. Goss, M. J. Rayson, P. R. Briddon, and J. M. Baker (2007) Metastable
       Frenkel pairs and the W11−W14 electron paramagnetic resonance centers in
       diamond Phys. Rev. B 76, 045203 https://doi.org/10.1103/PhysRevB.76.045203

[100] B.G. Svennson, C.Jagadish, A.Hallen, J.Lalita (1995) Point defect in MeV ion-
       implanted silicon studied by deep level transient spectroscopy Nucl. Instr. Meth.
       Phys. Research B106 183-190 https://doi.org/10.1016/0168-583X(95)00702-4

[101] V.V. Emtsev, T.V.Mashovets, V.V. Miknovich (1992) Frenkel pairs in germanium
       and silicon Soviet Physics- Semiconductors 24 12-25

[102] H. Bracht, N.A. Stolwijk, H.Meher (1995) Properties of intrinsic point defects in
       silicon determined by zinc diffusion measurements under non equilibrium
       conditions Phys.Rev.B 52 16542-16560
       https://doi.org/10.1103/PhysRevB.52.16542

[103] R. Habu, T. Iwasaki, H. Harada, and A. Tomiura (1994) Diffusion Behaviour of
       point defects in Si crystals during melt growth IV: numerical analysis Jpn. J. Appl.
       Phys. 33 1234-1242 https://doi.org/10.1143/JJAP.33.1234

[104] N. Fukata, A. Kasuya, A Suezawa (2001) Formation energy of vacancy in silicon
       determined by a new quenching method Physica B 308-310 1125-1128
       https://doi.org/10.1016/S0921-4526(01)00908-5

[105] T. Südkamp, H. Bracht (2016) Self-diffusion in crystalline silicon: A single
       diffusion activation enthalpy down to 755°C Phys.Rev.B 94 125208
       https://doi.org/10.1103/PhysRevB.94.125208

[106] E. Hüger, U. Tietze, D. Lott,H.Bracht, D.Bougerad, E.E.Haller, H.Schmidt (2008)
       Selfdiffusion in germanium isotope multilayers at low temperatures
       Appl.Phys.Lett. 93 162104 https://doi.org/10.1063/1.3002294

[107] M. Tang, L. Colombo, J. Zhu, and T. Diaz de la Rubia, (1997) Intrinsic point
       defects in crystalline silicon: Tight-binding molecular dynamics studies of self-
       diffusion, interstitial-vacancy recombination, and formation volumes Phys. Rev. B
       55 14 279 https://doi.org/10.1103/physrevb.55.14279

[108] P.E. Blöchl, E. Smargiassi, R. Car, D.B. Laks, W.Andreoni, S.T.Pantelides(1993) First principle calculations of self-diffusion constants in silicon Phys.Rev.Lett. 70 2435 https://doi.org/10.1103/PhysRevLett.70.2435

[109] M. Dionizio Moreira, R.H. Miwa, P. Venezuela (2004) Electronic and structural properties of selfinterstitials in germanium Phys.Rev.B 70 115215 https://doi.org/10.1103/PhysRevB.70.115215

[110] D. A. Drabold, S. K.E streicher (2007) Defect theory: An armchair Hystory in Theory of defects in semiconductors D.A.Drabold and Stefan K.Estreicher (Eds) pp.11-28 Springer Verlag, Berlin Heidelberg

[111] P. Spiewak, J. Vanhellemont, K.Sueoka, K. Kurzydlowski,I.Romandcic (2008)First principle calculations of formation energies and deep level associated with neutral and charged vacancy in germanium J.Appl. Phys.103 086103 https://doi.org/10.1063/1.2907730

[112] K. Estreicher, D. Backlund, and T.M. Gibbons, (2010) Theory of Defects in Si and Ge: past, present and recent developments Thin Solid Films 518, 2413-2417 https://doi.org/10.1016/j.tsf.2009.09.131

[113] P. Spiewak, J. Vanhellemont, K. Kurzydlowski (2011) Improved calculation of vacancy properties in Ge using the HSE range-separated hybrid functional J.Appl.Phys. 110 063534 https://doi.org/10.1063/1.3642953

[114] P. Spiewak, K. Kurzydlowski (2013) Formation and migration energies of the vacancy in silicon calculated using the HSE06 range-separated hybrid functional Phys.Rev. B 88 195204 https://doi.org/10.1103/PhysRevB.88.195204

[115] G. Yang , H. Mei, YT. Guan, GJ. Wang, DM. Mei, K. Irmscher (2015) Study on the Properties of High Purity Germanium Crystals J.Physics: Conference Series 606 012013 (8 p)

[116] T. Taishi, H. Ise, Y. Murao, T. Osawa, M. Suezawa, Y. Tokumoto, Y. Ohno, K. Hoshikawa, I. Yonen aga (2010) Czochralski-growth of germanium crystals containing high concentrations of oxygen impurities J. Crystal Growth 312 2783-2787 https://doi.org/10.1016/j.jcrysgro.2010.05.045

[117] F. C. Frank, D. Turnbull (1956) Mechanism of diffusion of copper in germanium Phys.Rev. 10 617-618 https://doi.org/10.1103/PhysRev.104.617

[118] U. Gösele, W. Frank, A.Seeger (1980) Mechanism and kinetics of the diffusion of gold in silicon Appl.Phys. 23 361-368 https://doi.org/10.1007/BF00903217

[119] H. Bracht, N.A. Stolwijk, H. Mehrer, I. Yonenaga (1991) Short-time diffusion of zinc in silicon for the study of intrinsic point defects Appl. Phys. Lett. 59 3559-1561 https://doi.org/10.1063/1.106393

[120]. M. Pesola, J. von Boehm, T. Mattila, R. M. Nieminen (1999) Computational study of interstitial oxygen and vacancy-oxygen complexes in silicon, Phys.Rev.B 60 11449- 11463 https://doi.org/10.1103/PhysRevB.60.11449

[121] J. Coutinho, R. Jones, P.R. Briddon, S.Öberg (2000) Oxygen and dioxygen centers in Si and Ge: Density-functional calculations Phys.Rev.B 62 10824 -10840 https://doi.org/10.1103/physrevb.62.10824

[122] R. A. Casali, H. Rücker, M. Methfessel (2001) Interaction of vacancies with interstitial oxygen in silicon Appl.Phys. Lett. 78 913 https://doi.org/10.1063/1.1347014

[123] M. Furuhashi, K. Taniguchi (2005) Diffusion and dissociation mechanisms of vacancy-oxygen complex in Silicon Appl. Phys. Lett. 86 142107

[124] D. J. Backlund, S. K. Estreicher (2010) Ti, Fe, and Ni in Si and their interactions with the vacancy and the A center: A theoretical study Phys. Rev. B 81 235213/1-8 https://doi.org/10.1103/PhysRevB.81.235213

[125] C.A. Londos, E.N.Sgourou, D.Hall, A.Chroneos (2014) Vacancy-oxygen defects in silicon: the impact of isovalent doping J.Mater.Sci. Mater.Electron 25 2395-2410 https://doi.org/10.1007/s10854-014-1947-6

[126] G. D. Watkins and J. W. Corbett, (1961) Defects in Irradiated Silicon. I. Electron Spin Resonance of the Si-A Center, Phys. Rev. 121 1001 https://doi.org/10.1103/PhysRev.121.1001

[127] J. W. Corbett, G. D. Watkins, R. M. Chrenko, R. G. McDonald, (1961) Defects in Irradiated Silicon. II. Infrared Absorption of the Si-A Center, Phys.Rev. 121 1015 https://doi.org/10.1103/PhysRev.121.1015

[128] R. E. Whan (1965) Evidence for low temperature motion of vacancies in germanium Appl. Phys. Lett. 6 221 https://doi.org/10.1063/1.1754143

[129] R. E. Whan (1965) Investigations of Oxygen-Defect Interactions between 25 and 700 K in Irradiated Germanium, Phys. Rev. A 140 690 https://doi.org/10.1103/PhysRev.140.A690

[130] V. V. Litvinov, L. I. Murin, J. L. Lindstrom, V. P. Markevich, and A. N. Petukh (2002) Local vibration modes of the oxygen-vacancy complex in germanium Semiconductors 36 621-624 https://doi.org/10.1134/1.1485658

[131] V. P. Markevich, A. R. Peaker, L. I. Murin, and N. V. Abrosimov (2003) Vacancy–oxygen complex in Si 1-x Ge x crystals Appl. Phys. Lett. 82 2652 https://doi.org/10.1063/1.1569422

[132] C. Londos, G. Antonaras, M. Potsidi, A. Misiuk, and V. Emtsev, (2005) The effect of thermal treatments on the annealing behaviour of oxygen-vacancy complexes in irradiated carbon doped silicon Solid State Phenomena, 108-109 205–210 https://doi.org/10.4028/www.scientific.net/SSP.108-109.205

[133] A. Chroneos, E. N. Sgourou, C. A. Londos, and U. Schwingenschl (2015) Oxygen defect processes in silicon and silicon germanium Appl.Phys. Rev. 2 021306

[134] M. F. Rye-Larsen (2016) Oxygen-related defects in Carbon-rich Solar Silicon studied by Fourier Transform Infrared Spectroscopy Master's Thesis May 2016 University of Oslo

[135] C. Claeys, (2008) High purity silicon ECS Transactions 16 97-108

[136] A. Chroneos, C. A. Londos, E.N. Sgourou (2014) Strategies to suppress A-center formation in silicon and germanium from a mass action analysis viewpoint physics.karazin.ua/doc/v_20_2014/pdf/1113/87-90.pdf ;

[137] A. Chroneos, C.A. Londos, E.N.Sgourou (2011) Effect of tin doping on oxygen- and carbon-related defects in Czochralski silicon, J.Appl.Phys. 110 093507 https://doi.org/10.1063/1.3658261

[138] L. C. Kimerling, M. T. Asom, J. L. Benton, P. J. Drevinsky, C. E. Caefer (1898) Interstitial Defect Reactions in Silicon, Mater. Sci. Forum 38-41 141 https://doi.org/10.4028/www.scientific.net/MSF.38-41.141

[139] A. Chroneos and C. A. Londos, (2010) Interaction of A-centers with isovalent impurities in silicon J. Appl. Phys. 107 093518 https://doi.org/10.1063/1.3409888

[140] V. P. Markevich, I. D. Hawkins, and A. R. Peaker, K. V. Emtsev and V. V. Emtsev (2004) Vacancy–group-V-impurity atom pairs in Ge crystals doped with P, As, Sb, and Bi Phys.Rev. B 70, 235213

[141] A. R. Peaker, V. P. Markevich, F. D. Auret, L. Dobaczewski and N.Abrosimov (2005) The vacancy–donor pair in unstrained silicon, germanium and SiGe alloys Journal of Physics: Condensed Matter 17 (22) S2293- S2302 https://doi.org/10.1088/0953-8984/17/22/018

[142] P. Rava, H.C.Gatos, J.Lagowski (1982) Thermally activated oxygen donors in Si J.Electrochem.Soc. 129 2844-2849 https://doi.org/10.1149/1.2123690

[143] A. Ourmazd, W. Schröter, A. Bourret (1984) Oxygen-related thermal donors in silicon: A new structural and kinetic model J. Appl. Phys. 56, 1670 https://doi.org/10.1063/1.334156

[144] [D.J. Chadi (1990) Oxygen-oxygen complexes and thermal donors in silicon Phys. Rev.B 41 10595; D. J. Chadi, (1996) Core Structure of Thermal Donors in Silicon, Phys. Rev. Lett., 77 861–864

[145] T. Gregorkievicz and H.H.P.Th Bekman (1989) Thermal donors and oxygen related complexes in silicon Mat.Sci.Eng. B4 291-297 https://doi.org/10.1016/0921-5107(89)90260-2

[146] A. Chroneos (2008) Defect processes in germanium, PhD Thesis , Imperial College of Science,Technology and Medicine, London (UK) January 2008

[147] G. Davies, A. T. Collins (1993) Vacancy complexes in diamond Diamond and Related Materials 2 (2–4) 80-86 https://doi.org/10.1016/0925-9635(93)90035-Z

[148] H. Lim, S. Parj, H. Cheong, H.-M. Choi, Y.- C. Kim (2006) Photoluminescence of Natural Diamonds J. Korean Phys. Soc. 48 155-1559

[149] P. Maletinsky, S. Hong, M. S. Grinolds, B. Hausmann, M. D. Lukin, R. L. Walsworth, M. Loncar, & A. Yacoby (2012) A robust scanning diamond sensor for nanoscale imaging with single nitrogen-vacancy centres Nature Nanotechnology 7, 320-324 https://doi.org/10.1038/nnano.2012.50

[150] T. Müller, C. Hepp, B. Pingault, E. Neu, S. Gsell, M. Schreck, H. Sternschulte, D. Steinmüller-Neth, C. Becher, M. Atatüre (2014) Optical signatures of silicon-vacancy spins in diamond Nature Comm 5 3328

[155] L. Thiel, D. Rohner, M. Ganzhorn, P. Appel, E. Neu, B. Müller, R. Kleiner, D. Koelle and P. Maletinsky (2016) Quantitative nanoscale vortex imaging using a cryogenic quantum magnetometer Nature Nanotechnology 11 677–681 https://doi.org/10.1038/nnano.2016.63

[156] M. Pelliccione, A. Jenkins, P.Ovartchaiyapong, Ch. Reetz, E. Emmanouilidou, Ni Ni, A. C. Bleszynski Jayich (2016) Scanned probe imaging of nanoscale magnetism at cryogenic temperatures with a single-spin quantum sensor Nature Nanotechnology 11 700-705 https://doi.org/10.1038/nnano.2016.68

[157] P. Deák, B. Aradi, M. Kaviani, T. Frauenheim and Adam Gali (2014) The formation of NV centers in diamond: A theoretical study based on calculated transitions and migration of nitrogen and vacancy related defects. Phys.Rev B. 89 075203 https://doi.org/10.1103/PhysRevB.89.075203

[158] J. Wrachtrup, F.Jelexko, B.Grotz, L.McGuiness (2013) Nitrogen –vacancy centers close to surfaces MRS Bull. 38 149-154 https://doi.org/10.1557/mrs.2013.22

[159] J. Schwartz, P Michaelides, C D Weis and T. Schenkel (2011) In situ optimization of co-implantation and substrate temperature conditions for nitrogen-vacancy center formation in single-crystal diamonds New J. Physics, 13 035022

[160] R. Schirhagl, K. Chang, M. Loretz, C.L.Degen (2014) Nitrogen-vacancy centers in diamond: Nanoscale sensors for physics and biology Ann.Rev.Phys.Chem. 65 83-105 https://doi.org/10.1146/annurev-physchem-040513-103659

[161] M.W. Doherty, B. Manson, P. Delaney, L. C. L. Hollemberg (2011) The negatively charged nitrogen-vacancy centre in diamond: the electronic solution New J. Phys. 13 025019 https://doi.org/10.1088/1367-2630/13/2/025019

[162] L. V.C. Assali, R. Larico, W.V.M. Machado, J. F. Justo (2005) Nickel-Vacancy Complexes in Diamond: An Ab-Initio Investigation Materials Science Forum 483-485 1043-1046 https://doi.org/10.4028/www.scientific.net/MSF.483-485.1043

[163] J. Jeleko J.F. Wrachtrup (2006) Single defects in diamond: a Review Phys.Stat. Sol.A 203 3207-3225 https://doi.org/10.1002/pssa.200671403

[164] S. Pezzagna, D. Rogalla, D. Wildanger, J. Meijer and . Zaitsev (2011) Creation and nature of optical centres in diamond for single-photon emission—overview and critical remarks New Journal of Physics 13 035024 (27pp)

[165] T.Iwasaki, F. Ishibashi,Y. Miyamoto,Y. Doi, S. Kobayashi T. Miyazaki, K. Tahara, K. D. Jahnke, L. J. Rogers, B. Naydenov, F. Jelezko, S. Yamasaki, S. Nagamachi, T. Inubushi, N Mizuochi, and M. Hatano (2015) Germanium-Vacancy Single Color Centers in Diamond Nature Scientific Reports 5.12882 /DOI 10.1038/srep12882

[166] J. Ruan, W. Choyke, and K. Kobashi (1993) Oxygen-related centers in chemical vapor deposition of diamond Appl. Phys. Lett. 62 1379 https://doi.org/10.1063/1.108685

[167] Y.G. Zhang, Z. Tang, X.G. Zhao, J. Chu (2014) A neutral oxygen-vacancy centre in diamond: A plausible qubit candidate and its spintronic and electronic properties Appl.Phys. Letters 105 052107-052107-4 https://doi.org/10.1063/1.4892654

[168] A Gali, J E Lowther and P Deák (2001) Defect states of substitutional oxygen in diamond J. Phys.: Condens. Matt. 13 11607-11613 https://doi.org/10.1088/0953-8984/13/50/319

[169] I. Aharonovich (2009) Two level ultrabright single photon emission from diamond nanocrystals Nano Lett. 9 3191-3195 https://doi.org/10.1021/nl9014167

[170]C. Glover, M.E. Newton , P. Martineau , D.Twitchen and J.M. Baker (2003) Hydrogen Incorporation in Diamond: The Nitrogen-Vacancy-Hydrogen Complex Phys. Rev. Lett. 90 185507 https://doi.org/10.1103/PhysRevLett.90.185507

[171] J. P. Goss, P R Briddon, R Jones and S. Sque (2003) The vacancy–nitrogen–hydrogen complex in diamond: a potential deep centre in chemical vapour deposited material J. Physics: Condensed Matter 15 S2903–S2911 https://doi.org/10.1088/0953-8984/15/39/014

[172] A. Scholze, W. G. Schmidt,and F. Bechstedt (1966) Structure of the diamond 111 surface: Single-dangling-bond versus triple-dangling-bond face Phys.Rev B 53
[173] G. J. Dienes, D. O Welch (1987) Comment on "Diffusion without vacancies or interstitials: A new concerted exchange Mechanism Phys.Rev.Lett. 59 843

[174] K. C. Pandey (1986) Diffusion without Vacancies or Interstitials: A New Concerted Exchange Mechanism Phys.Rev. Lett. 57 2287 https://doi.org/10.1103/PhysRevLett.57.2287

[175] A. Ural, P. B. Griffin, and J. D. Plummer (1998) Experimental evidence for a dual vacancy–interstitial mechanism of self-diffusion in silicon Appl. Phys. Lett. 73 1706 https://doi.org/10.1063/1.122252

[176] D. Caliste, P. Pochet, T. Deutsch, and F. Lançon (2007) Germanium diffusion mechanisms in silicon from first principles Phys.Rev B. 75, 125203 https://doi.org/10.1103/PhysRevB.75.125203

[177] Yuan-Ting Liao (2010) Diffusion in SiGe and Ge, PhD Dissertation, University of California-Berkeley

[178] K. T. Koga, M. J. Walter, E. Nakamura, and K. Kobayashi (2005), Carbon self-diffusion in a natural diamond, Phys. Rev. B 72 024108 https://doi.org/10.1103/PhysRevB.72.024108

[179] B. Zhang, X.Wu (2012) Calculation of self-diffusion coefficients in diamond Appl. Phys. Lett. 100 051901 https://doi.org/10.1063/1.3680600

[180] Y .Shimizu, M. Uematsu, and K. M. Itoh (2007) Experimental evidence of Mediated Silicon Self-Diffusion in Single-Crystalline Silicon Phys. Rev. Lett. 98 095901 https://doi.org/10.1103/PhysRevLett.98.095901

[181] R. Kube and H. Bracht ,E. Hüger, H. Schmidt, J. W. Ager III, E. E. Haller, T. Geue, J. Stahn (2013) Contributions of vacancies and self-interstitials to self-

diffusion in silicon under thermal equilibrium and non-equilibrium conditions Phys.Rev B 88 085206 https://doi.org/10.1103/PhysRevB.88.085206

[182] R. N. Hall and J. H. Racette (1964)Diffusion and Solubility of Copper in Extrinsic and Intrinsic Germanium, Silicon, and Gallium Arsenide J. Appl. Phys. 35 379-396 https://doi.org/10.1063/1.1713322

[183] S.J. Breuer, P.R. Briddon (1995) Ab initio investigation of the native defects in diamond and self-diffusion. Phys Rev B 51 6984-6994 https://doi.org/10.1103/PhysRevB.51.6984

[184] H. Bracht, E.E. Haller, K.Eber, M.Cardona, R.Clark-Phelps (1998) Self-diffusion in isotopically controlled heterostructures of elemental and compound semiconductors MRS Symp.Proc. Vol.527 335-346

[185] A Gösele, A. Plössl, and T. Y. Tan (1996) in Process Physics and Modeling in Semiconductor Technology G. R. Srinivasan, C. S. Murthy, and S. T. Dunham Edt. p. 309 Electrochemical Society,Pennington, NJ,

[186] G Vogel, G Hettich and H Mehrer (1983) Self-diffusion in intrinsic germanium and effects of doping on self-diffusion in germanium, J. of Physics C: Solid State Physics, 16, 6197-6204 https://doi.org/10.1088/0022-3719/16/32/012

[187] A. Chroneos,H. Bracht, R. W. Grimes, and B. P. Uberuaga (2008) Vacancy-mediated dopant diffusion activation enthalpies for germanium, Appl.Phys.Lett. 92 172103 https://doi.org/10.1063/1.2918842

[188] T.Y.Tan, U.Gosele (1985) Point defects, diffusion processes, and swirl defect formation in silicon Appl.Phys.A 37 1-17 https://doi.org/10.1007/BF00617863

[189] H. Zimmermann, H.Ryssel (1992) The Modeling of Pt diffusion in silicon under non-equilibrium conditions J.Electrochem.Soc. 139 256-262 https://doi.org/10.1149/1.2069180

[190] H. Zimmermann (1998) Vacancy distribution in silicon and methods for their accurate determination Defects and diffusion Forum 153-155 111-136

[191] M. Jacob, P. Pichler, H. Ryssel, R. Falster (1997) Determination of vacancy concentrations in the bulk of silicon wafers by platinum diffusion experiments J. Appl. Phys. 82 182-191 https://doi.org/10.1063/1.365796

[192] S. Obeidi, N.A.Stolwijk (2001) Diffusion of iridium in silicon: changeover from a foreign-atom limited to a native-defect cotrolled transport mode Phys.Rev.B 64 113201-1-4 https://doi.org/10.1103/physrevb.64.113201

[193] L. Lerner and N. A. Stolwijk (2005) Vacancy concentrations in silicon determined by the in-diffusion of iridium Appl. Phys. Lett 86 011901 https://doi.org/10.1063/1.1844031

[194] S. Dannefaer, P. Mascher, D. Kerr (1986) Monovalence formation enthalpy in silicon Phys. Rev.Lett. 56 2195-219 https://doi.org/10.1103/PhysRevLett.56.2195

[195] M. O. Henry, J.D.Campion, K.G. McGuigiant, E.C.Lightowlers, M.C.do Carmo,M.H.Nazare (1994) A photoluminescence study of ZnO complexes in silicon Semicond.Sci.Technol. 9 1375-1381 https://doi.org/10.1088/0268-1242/9/7/014

[196] A. Chroneos (2008) Stability of impurity-vacancy pairs in germanium carbide J.Mat-Sci:Mater-Electron. 19 25-28 https://doi.org/10.1007/s10854-007-9132-9

[197] A. Mattausch (2005) Ab initio-Theory of Point Defects and Defect Complexes in SiC,Thesis Friedrich-Alexander-Universität Erlangen-Nürnberg 7/7/2005

[198] D. P. Birnie III, W. D. Kingery (1990) The limit of non-stoichiometry in silicon carbide J.Mater.Science 25 2827-2834 https://doi.org/10.1007/BF00584888

[199] J.D. Hong, M.H. Hon, R.F. Davis (1979) Self-diffusion in alpha and beta silicon carbide Ceramurgia International 5 4155-41160 https://doi.org/10.1016/0390-5519(79)90024-3

[200] R. N. Ghoshtagore and R. L. Coble (1966) Self-Diffusion in Silicon Carbide Phys. Rev. 143, 623 – 626

[201] M. K. Linnarsson, M. S. Janson, J. Zhang, E. Janzén and B. G. Svensson (2004) Self-diffusion of 12C and 13C in intrinsic 4H–SiC J. Appl. Phys. 95 8469-8471 https://doi.org/10.1063/1.1751229

[202] K. Rüschenschmidt, H. Bracht, N. A. Stolwijk, M. Laube, G. Pensl and G. R. Brandes (2004) Self-diffusion in isotopically enriched silicon carbide and its correlation with dopant diffusion J. Appl. Phys. 96 1458-1463 https://doi.org/10.1063/1.1766101

[203] M. Bockstedte, A. Gali, A. Mattausch, O. Pankratov, J. W. Steeds (2008) Identification of intrinsic defects in SiC: Towards an understanding of defect aggregates by combining theoretical and experimental approaches phys.stat.sol.B 245 1282-1297 https://doi.org/10.1002/pssb.200844048

[204] A. Kawasuso, M.Weidner, F.Redmann,T.Frank, P.Sperr,G.Kögel, M.Yoshikava, H.Itoh,R-Krause-Rehber, W.Triftshäuser, G.Pensl (2004) Vacancy defects

detected by positron annhihilation in Silicon Carbide, Recent Major Advances, W.J.Choyke,H.Matsunami,G.Pensl(Edt.) pp.563-584

[205] F. Redmann, A. Kawasuso, K. Petters, H. Itoh, R. Krause-Rehberg (2001) Illumination effects in irradiated 6H n-type SiC observed by positron annihilation spectroscopy Physica B 308–310 629–632 https://doi.org/10.1016/S0921-4526(01)00764-5

[206] T. Son, M. Wagner, C.G. Hemmingsson, L.Storasta, B. Magnusson, W.M.Chen, S. Greulich-Weber, J.M.Spaeth, J.Janzen (2004) Electronic structure of deep defects inSiC in Silicon Carbide, Recent Major Advances, W.J.Choyke,H.Matsunami,G.Pensl (Edt.) pp.461-492]   Springer Verlag, Berlin Heidelberg

[207] L Torpo, M Marlo, T E M Staab and R M Nieminen (2001) Comprehensive ab initio study of properties of monovacancies and antisites in 4H-SiC J.Phys.Condens.Matter 13 6203-6231 https://doi.org/10.1088/0953-8984/13/28/305

[208] A. Galeckas, J. Linnros, B. Breitholtz, and H. Bleichner (2001) Application of optical emission microscopy for reliability studies in 4H–SiC p + /n - /n + diodes J. Appl. Phys. 90 980-984 https://doi.org/10.1063/1.1380221

[209] C. Wang, J. Bernholc, R. F. Davis (1988) Formation energies, abundances, and the electronic structure of native defects in cubic SiC Phys. Rev.B 38 12752 -12755 https://doi.org/10.1103/PhysRevB.38.12752

[210] A. Gali, P. Deák, P. Ordejón, N. T. Son, E. Janzén, W. J. Choyke (1998) Aggregation of carbon interstitials in silicon carbide: A theoretical study Phys. Rev.B 57 125201

[211] L. Torpo S. Pöykkö, and R.M. Nieminen (1998) Antisites in silicon carbide Phys. Rev.B 57 6243-6246 https://doi.org/10.1103/PhysRevB.57.6243

[212] A. Zywietz, J. Furthmüller, F. Bechstedt (1999) Vacancies in SiC: Influence of Jahn-Teller distortions, spin effects, and crystal structure Phys.Rev.B 59 15166-15180 https://doi.org/10.1103/physrevb.59.15166

[213] M. Bockstedte, A. Mattausch, O. Pankratov (2003) Ab initio study of the migration of intrinsic defects in 3C-SiC Phys .Rev.B 68 205201(1-17)

[214] J. M. Lento, L. Torpo, T. E. M. Staab, R. M. Nieminen (2004) Self-interstitials in 3C-SiC J.Phys. Condens.Matter 16 1053-1060 https://doi.org/10.1088/0953-8984/16/7/005

[215] F. Bernardini, A. Mattoni, L. Colombo (2004) Energetics of native point defects in cubic silicon carbide Eur. Phys.J B 38 437-444 https://doi.org/10.1140/epjb/e2004-00137-6

[216] F. Bruneval, G. Roma (2011) Energetics and metastability of the silicon vacancy in cubic SiC Phys. Rev.B 83 144 116

[217] E. Rauls, T. Lingner, Z. Hajnal, S. Greulich-Weber, T. Frauenheim, J. M. Spaeth (2000) Metastability of the Neutral Silicon Vacancy in 4H-SiC phys. stat. sol. (b) 217, R1-R3

[218] M. Bockstedte, A. Mattausch, O. Pankratov (2004) Defect migration and annealing mechanism in Silicon carbide.Recent mayor advances W.J.Choyke, H.Marsunami, G.Pensl (Eds) pp.27-55 Springer- Verlag Berlin, Heidelberg New York

[219] M. Bockstedte, A. Mattausch, and O. Pankratov, (2004) Ab initio study of the annealing of vacancies and interstitials in cubic SiC: vacancy-interstitial recombination and aggregation of carbon interstitials, Phys. Rev. B 69, 235202–13 https://doi.org/10.1103/PhysRevB.69.235202

[220] G. Roma, F. Bruneval, T. Liao, O. N. Bedoya Martínez, and J-P.Crocombette (2012) Formation and migration energy of native defects in silicon carbide from first principles: an overview Defect and Diffusion Forum 323-325 11-18

[221] F. Bechstedt, J. Furtmüller, U.Grossner, C.Raffy (2004) Zero-and two-dimensional native defects in Silicon carbide. Recent mayor advances W.J.Choyke, H.Marsunami, G.Pensl (Eds) pp.3-25 Springer- Verlag Berlin, Heidelberg New York

[222] A. Le Donne, S. Binetti, M. Acciarri, S. Pizzini (2004) Electrical characterization of electron irradiated X-rays detectors based on 4H-SiC epitaxial layers Diamond and Related Materials 13 414–418 https://doi.org/10.1016/j.diamond.2003.11.079

[223] A. LeDonne,S. Binetti, S. Pizzini (2005) Electrical and optical characterization of electron-irradiated 4H-SiC epitaxial layers annealed at low temperature Diamond and related materials 14 1150-1153 https://doi.org/10.1016/j.diamond.2004.10.020

[224] S. G. Sridhara, L. L. Clemen, R. P. Devaty, W. J. Choyke, D. J. Larkin, H. S. Kong, T. Troffer and G. Pensl ((1998) Photoluminescence and transport studies of boron in 4H SiC J. Appl. Phys. 83 7909-7919 https://doi.org/10.1063/1.367970

[225] J.D. Hong, R. F. Davis, D. E. Newbury (1981) Self-diffusion of silicon-30 in a-SiC single crystals J. Mater. Science 16 2485–2494 https://doi.org/10.1007/BF01113585

[226] J. D. Hong, R. F. Davis(1980) Self-Diffusion of Carbon-14 in High-Purity and N-Doped a-SiC Single Crystals J.Am.Ceram.Soc.63 546-552 https://doi.org/10.1111/j.1151-2916.1980.tb10762.x

[227] J.D. Hong, M.H. Hon, R.F. Davis (1979) Self-diffusion in alpha and beta silicon carbide Ceramurgia International 5 4155-41160 https://doi.org/10.1016/0390-5519(79)90024-3

[228] T.Y.Tan, U.Gosele (1985) Point defects, diffusion processes, and swirl defect formation in silicon Appl.Phys.A 37 1-17 https://doi.org/10.1007/BF00617863

[229] A. Borghesi, B.Pivac, A.Sassella, A,Stella (1995) Oxygen precipitation in silicon J.Appl.Phys. 77 4169-4244 https://doi.org/10.1063/1.359479

Keyword Index

About the Author

Sergio Pizzini is a former full professor of Physical Chemistry at the Department of Materials Science, University of Milano-Bicocca. He started his scientific carrier at the Joint Research Centre of the European Commission in Ispra (Italy) and later in Petten (Nederland), where he was committed of thermodynamic and electrochemical studies on molten fluorides and ionic oxides, in cooperation with the Oak Ridge nuclear center in USA. A notable by-product of the electrochemical studies was the development of a reversible hydrogen electrode in molten KHF_2.

After leaving the Commission, he joined the University of Milano, where he started basic and applied studies on solid oxide electrolytes, addressed at the development of solid state sensors, with some patent applications. Still maintaining his position at the University, he worked for four years as Director of the Materials Department of the Corporate Research Centre of Montedison in Novara, where he supervised new R&D activities on advanced materials for electronics, including InP and silicon. In the next few years, as the CEO of Heliosil, a Research Company sponsored by italian metallurgical silicon producers and, later, by ENI, the major Italian Oil Company, he developed and patented a new process for the production of solar grade silicon and a new process for the directional solidification of silicon, starting also systematic studies on extended defects in silicon.

In 1982 he left any outside duty for serving the University of Milano, and later, the University of Milano-Bicocca. There, his main research interests have been in the areas of structural, electronic and optical properties of point defects, extended defects and impurities in single crystal and multicrystalline silicon, with major emphasis on grain boundaries, dislocations and oxygen and carbon impurities, in the frame of national and European Projects, as local or European coordinator and of joint R&D activities with major italian Companies operating in the sector.

After retirement, he continued the cooperation with the University of Milano-Bicocca on a voluntary basis.He was Chairman or Co-Chairman of a number of International Symposia in the Materials Science field in Europe, USA and China, including the last EMRS 2008, 2010, 2012 and 2014 Symposia on Advanced Silicon and Silicon-Germanium Materials.

He is author of more than 250 technical papers published in peer reviewed international Journals, and authored or co-authored five books, two under Wiley &Sons and two under CRC Press.